IMPA Monographs

Volume 2

This series, jointly established by IMPA and Springer, publishes advanced monographs giving authoritative accounts of current research in any field of mathematics, with emphasis on those fields that are closer to the areas currently supported at IMPA. The series gives well-written presentations of the "state-of-the-art" in fields of mathematical research and pointers to future directions of research.

Series Editors

Emanuel Carneiro, *Instituto de Matemática Pura e Aplicada*
Severino Collier, *Universidade Federal do Rio de Janeiro*
Claudio Landim, *Instituto de Matemática Pura e Aplicada*
Paulo Sad, *Instituto de Matemática Pura e Aplicada*

More information about this series at http://www.springer.com/series/13501

Jorge Vitório Pereira • Luc Pirio

An Invitation to Web Geometry

 Springer

Jorge Vitório Pereira
Instituto de Matemática Pura e Aplicada
Rio de Janeiro, Brazil

Luc Pirio
Institut de Recherches Mathématiques
  de Rennes
IRMAR, UMR 6625 du CNRS
Université Rennes 1, Rennes
France

IMPA Monographs
ISBN 978-3-319-38508-2   ISBN 978-3-319-14562-4   (eBook)
DOI 10.1007/978-3-319-14562-4

Mathematics Subject Classification (2010): 53A60, 53A20, 14H

Springer Cham Heidelberg New York Dordrecht London
© Springer International Publishing Switzerland 2015
Softcover reprint of the hardcover 1st edition 2015

Printed on acid-free paper

Springer International Publishing AG Switzerland is part of Springer Science+Business Media
(www.springer.com)

Para Dayse e Jorge,
os meus Pastores.
J.V.P.

Pour Min' et Nin'
L.P.

Preface

The first purpose of this text was to serve as supporting material for a minicourse on web geometry delivered at the 27th Brazilian Mathematical Colloquium, which took place at IMPA in the last week of July 2009. Still, it contains much more than what can possibly be covered in five lectures of 1 h each. The abundance of material is due to the second purpose of this text: to convey some of the beauty of web geometry and provide an account, as self-contained as possible, of some of the exciting advancements the field has witnessed in the last few years.

We have tried to make this text as little demanding as possible in terms of prerequisites.

It is true that familiarity with the basic language of algebraic and complex geometry is sometimes welcome but, except in very few passages, not much more is needed. An effort has been made to explain, even if sometimes superficially, every unusual concept appearing in the text.

Contents of the Chapters

The table of contents tells rather precisely what the book is about. The following descriptions give additional information.

Chapter 1 is introductory and describes the basic notions of web geometry. Most of the content of this chapter is well known. A notable exception is the notion of duality for global webs on projective spaces $\mathbb{P}^n$, which appears to be new when $n > 2$. This notion is discussed in Sect. 1.4.3. The first two sections, more specifically Sects. 1.1 and 1.2, are of rather elementary nature and might be read by an undergraduate student.

Chapter 2 is about the notions of abelian relation and rank. It offers an outline of Abel's method to determine the abelian relations of a given planar web. It also gives a description of the abelian relations of planar webs admitting an infinitesimal symmetry. The most important results in this chapter are Chern's bound on the rank (Theorem 2.3.8) and the normal form for the conormals of a web of maximal rank

(Proposition 2.4.10). This last result is proved using a geometric approach based on classical concepts and results from projective algebraic geometry, which are described in detail.

Chapter 3 is devoted to Abel's notorious addition theorem. It first deals with the case of smooth projective curves, then tackles the general case after introducing the notion of abelian differential. Section 3.3 gives a rather precise description of the Castelnuovo curves (projective curves of maximal genus) hence of some algebraic webs of maximal rank. Section 3.4 expounds new results: an (easy) variant of Abel's theorem (Proposition 3.4.1), which is combined with Chern's bound on the rank so as to obtain bounds on the genus of curves included in abelian varieties (*cf.* Theorem 3.4.5).

Chapter 4 is where the converse to Abel's theorem is demonstrated. Its proof is given through a reduction to the planar case, which is then treated using a classical argument that can be traced back to Darboux. Then follows a presentation of some algebraization results. Important concepts such as Poincaré's and canonical maps for webs are discussed in this chapter. Our only contribution is of formal nature and is situated in Sect. 4.3, where we endeavor to work as intrinsically as possible.

Chapter 5 is entirely devoted to Trépreau's algebraization result. The proof that is presented is essentially the same as the original one [130]. The only "novelty" in this chapter is Sect. 5.1.2, where a geometric interpretation of the proof is given. As in the preceding chapter, an effort was made to formulate some of the results and their proofs as intrinsically as possible.

Chapter 6 takes up the case of planar webs of maximal rank, more specifically of exceptional planar webs. Classical criteria that characterize linearizable webs on the one hand and maximal rank webs on the other hand are explained. Then the existence of exceptional planar k-webs, for arbitrary $k \geq 5$, is established through the study of webs admitting infinitesimal automorphisms. The classification of the so-called CDQL webs on compact complex surfaces obtained recently by the authors is also reviewed. The chapter ends with a brief discussion about all the examples of planar exceptional webs we are aware of.

Finally, in the **Appendix**, the reader will find historical notes on the development of the field from its origins to recent advances.

How to Use This Book

The logical organization of this book is rather simple: the reader with enough time to spare can read it from cover to cover.

Those who are mostly interested in Bol–Trépreau's algebraization theorem may find useful the minimal route suggested by the graph below.

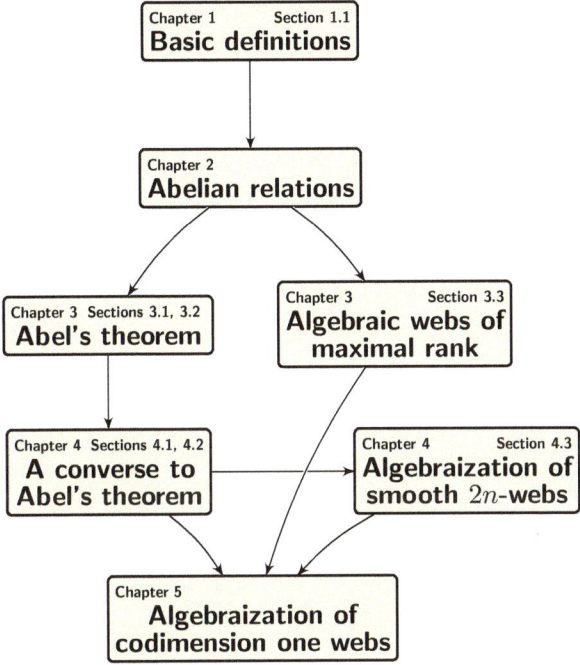

Those who are anxious to learn more about exceptional webs might prefer to use instead the following graph as a reading guide.

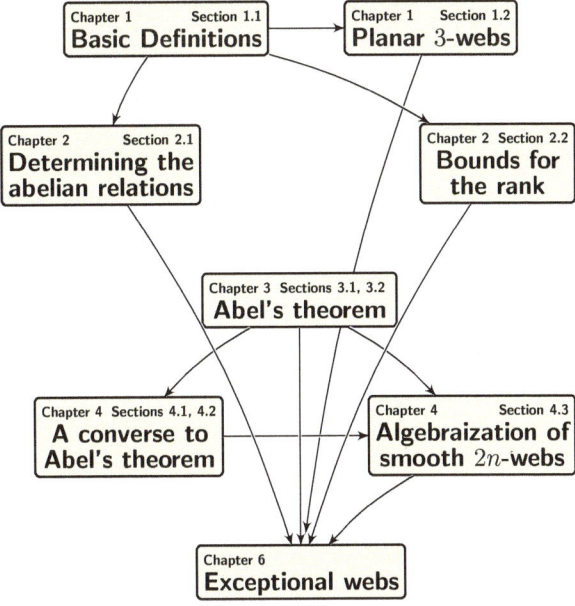

This book would have taken much longer to come to light without the invitation of Márcio Gomes Soares to submit a minicourse proposal to the 27th Brazilian Mathematical Colloquium. Besides Soares, we would like to thank Hernan Maycol Falla Luza and Paulo Sad, who caught a number of misprints and mistakes appearing in preliminary versions.

We are also indebted to Annie Bruter for her many corrections as well as for her help in translating into English a draft of the appendix originally written in French.

Last but not least, Jorge wishes to thank Dayse and Luc wishes to thank Mina for their patience and unconditional support during the writing of this book.

Rio de Janeiro, Brazil   Jorge Vitório Pereira
Rennes, France   Luc Pirio

Contents

List of Figures

Conventions

All the **definitions**, including this one, are presented in bold case and have a corresponding entry at the index.

Unless stated otherwise all the geometric entities like curves, surfaces, varieties, and manifolds considered in this text are reduced and complex holomorphic. Curves, surfaces, and varieties may be singular and may have several irreducible components. The manifolds are smooth connected varieties.

Web geometry lies on the interface of local differential geometry and projective algebraic geometry. Throughout the text, the reader will be confronted with both local non-algebraic subvarieties of the projective space and global, and algebraic and compact, projective subvarieties. A projective curve, surface, variety, or manifold will mean a compact curve, surface, variety, or manifold contained in some projective space. Beware that some authors use the term projective to qualify any subvariety, compact or not, algebraic or not, of a given projective space.

Throughout the text, there will be references to points $x \in (\mathbb{C}^n, 0)$ and properties of germs at the point x. The latter has to be understood as a point in a sufficiently small neighborhood of the origin and the property as a property of some representative of the germ defined in this very same sufficiently small neighborhood.

Below is a list of some of the notations used in this book.

- If n is a positive integer, $\underline{n}$ will stand for the set $\{1, \ldots, n\}$;
- For any $q \in \mathbb{N}$, $\mathbb{C}_q[x_1, \ldots, x_n]$ will stand for the vector space of degree q homogeneous polynomials in $x_1, \ldots, x_n$;
- The span of a subset S of a projective space or of a vector space will be denoted by $\langle S \rangle$;
- If f is a differentiable function depending on some variables $x_1, \ldots, x_n$, we will sometimes denote by f_{x_i} the partial derivative $\partial f / \partial x_i$.

Chapter 1
Local and Global Webs

In its classical form, web geometry studies local configurations of finitely many smooth foliations in general position. In Sect. 1.1 the basic definitions of our subject are laid down and the algebraic webs are introduced. These are among the most important examples of the whole theory.

Germs of webs defined by few foliations in general position are not particularly interesting. Basic results from differential calculus imply that the theory is locally trivial in this case. But as soon as the number of foliations surpasses the dimension of the ambient manifold, this is no longer true. The discovery of the curvature for 3-webs on surfaces during the last years of the 1920s is considered as marking the birth of web geometry. In Sect. 1.2 this curvature form is discussed and an early emblematic result of the theory that characterizes its vanishing is presented.

Although the emphasis of the theory is local, the most important examples are indeed globally defined on projective manifolds. In Sect. 1.3 the basic definitions are extended to encompass both germs of singular and global webs. Certainly more demanding than the previous sections, Sect. 1.3 should be read in parallel with Sect. 1.4 where the algebraic webs are revisited from a global viewpoint and is discussed how one can associate webs to linear systems on surfaces.

1.1 Basic Definitions

1.1.1 Germs of Smooth Webs

A **germ of smooth codimension one k-web** on $(\mathbb{C}^n, 0)$ is a collection

$$\mathcal{W} = \mathcal{F}_1 \boxtimes \cdots \boxtimes \mathcal{F}_k$$

© Springer International Publishing Switzerland 2015
J.V. Pereira, L. Pirio, *An Invitation to Web Geometry*, IMPA Monographs 2,
DOI 10.1007/978-3-319-14562-4_1

of k germs of smooth codimension one holomorphic foliations $\mathcal{F}_1, \ldots, \mathcal{F}_k$ such that their tangent spaces at the origin are in **general position**, that is, for any number $m \leq n$ of these foliations, the corresponding tangent spaces at the origin have an intersection of codimension m.

Usually the foliations $\mathcal{F}_i$ are presented by germs of holomorphic 1-forms $\omega_i \in \Omega^1(\mathbb{C}^n, 0)$, non-zero at $0 \in \mathbb{C}^n$ and satisfying Frobenius integrability condition $\omega_i \wedge d\omega_i = 0$. To present a germ of smooth web and keep track of its defining 1-forms, two alternative notations will be used:

$$\mathcal{W} = \mathcal{W}(\omega_1, \ldots, \omega_k) \qquad \text{or} \qquad \mathcal{W} = \mathcal{W}(\omega_1 \cdot \omega_2 \cdot \cdots \cdot \omega_k).$$

While the former is self-explanatory, the latter presents $\mathcal{W}$ as an unordered collection of k foliations defined by an element of $\mathrm{Sym}^k \Omega^1(\mathbb{C}^n, 0)$ (where $\cdot$ stands for the symmetric product of differential forms).

Notice that the general position assumption translates into

$$(\omega_{i_1} \wedge \cdots \wedge \omega_{i_m})(0) \neq 0$$

where $\{i_1, \ldots, i_m\}$ is any subset of $\underline{k}$ of cardinality $m \leq \min\{k, n\}$.

Since the foliations $\mathcal{F}_i$ are smooth they can be defined by level sets of submersions $u_i : (\mathbb{C}^n, 0) \to \mathbb{C}$. When profitable to present the web in terms of its defining submersions, the following notation will be used:

$$\mathcal{W} = \mathcal{W}(u_1, \ldots, u_k).$$

The germs of **quasi-smooth** webs on $(\mathbb{C}^n, 0)$ are defined by replacing the general position hypothesis on the tangent spaces at the origin by the weaker condition of pairwise transversality. Explicitly, a germ of quasi-smooth k-web $\mathcal{W} = \mathcal{F}_1 \boxtimes \cdots \boxtimes \mathcal{F}_k$ on $(\mathbb{C}^n, 0)$ is a collection of smooth foliations such that $T_0\mathcal{F}_i \neq T_0\mathcal{F}_j$ whenever i and j are distinct elements of $\underline{k}$.

There are similar definitions for webs of arbitrary (and even mixed) codimensions. Although extremely rich, the theory of webs of arbitrary codimension will not be discussed in this book (the interested reader can consult [3, 34, 59, 66] among many other references).

It is also interesting to study webs in different categories. For instance, one can paraphrase the definitions above to obtain differentiable, formal, ... webs. This book, unless stated otherwise, will stick to the holomorphic category.

1.1.2 Equivalence and First Examples

Local web geometry is ultimately interested in the classification of germs of smooth webs up to the natural action of $\mathrm{Diff}(\mathbb{C}^n, 0)$—the group of germs of biholomorphisms of $(\mathbb{C}^n, 0)$.

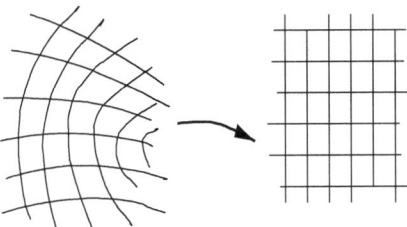

Fig. 1.1 There is only one smooth planar 2-web up to equivalence

If $\varphi \in \mathrm{Diff}(\mathbb{C}^n, 0)$ is a germ of biholomorphism, then the natural action just referred to is given by

$$\varphi^* \mathcal{W}\big(\omega_1 \cdot \omega_2 \cdot \,\cdots \cdot \omega_k\big) = \mathcal{W}\big(\varphi^*(\omega_1 \cdot \omega_2 \cdot \,\cdots \cdot \omega_k)\big) \,.$$

The germs of k-webs $\mathcal{W}(\omega_1 \cdot \omega_2 \cdot \,\cdots \cdot \omega_k)$ and $\mathcal{W}(\omega_1' \cdot \omega_2' \cdot \,\cdots \cdot \omega_k')$ will be considered **biholomorphically equivalent** if

$$\varphi^* \big(\omega_1 \cdot \omega_2 \cdot \,\cdots \cdot \omega_k\big) = u \cdot \big(\omega_1' \cdot \omega_2' \cdot \,\cdots \cdot \omega_k'\big)$$

for some germ of biholomorphism φ and some germ of invertible function $u \in \mathcal{O}^*_{(\mathbb{C}^n, 0)}$. In other words, there exists a permutation $\sigma \in \mathfrak{S}_k$—the symmetric group on k elements—such that the germs of 2-forms $\varphi^* \omega_i \wedge \omega_{\sigma(i)}'$ are identically zero for every $i \in \underline{k}$.

Clearly the biholomorphic equivalence defines an equivalence relation on the set of smooth k-webs on $(\mathbb{C}^n, 0)$. When the dimension of the space is greater than or equal to the number of defining foliations, that is when $n \geq k$, there is just one equivalence class (see Fig. 1.1 above). Indeed, if one considers a smooth k-web defined by k submersions $u_i : (\mathbb{C}^n, 0) \to \mathbb{C}$, then the map $U = (u_1, \ldots u_k) : (\mathbb{C}^n, 0) \to \mathbb{C}^k$ is a submersion thanks to the general position hypothesis. The constant rank Theorem ensures the existence of a biholomorphism $\varphi : (\mathbb{C}^n, 0) \to (\mathbb{C}^n, 0)$ taking the function u_i to the coordinate function x_i for every $i \in \underline{k}$. Symbolically, $\varphi^* u_i = x_i$.

When the number of defining foliations exceeds the dimension of the space by at least two ($k \geq n + 2$), there exists a multitude of equivalence classes through the following considerations. For a k-web $\mathcal{W} = \mathcal{F}_1 \boxtimes \cdots \boxtimes \mathcal{F}_k$, the tangent spaces of the foliations $\mathcal{F}_i$ at the origin determine a collection of k unordered points in $\mathbb{P}T_0^*(\mathbb{C}^n, 0) = \mathbb{P}^{n-1}$. The set of isomorphism classes of k unordered points in general position in a projective space $\mathbb{P}^{n-1}$ is the quotient of the open subset V of $(\mathbb{P}^{n-1})^k$ parametrizing k distinct points in general position by the action

$$\big((\sigma, g), (x_1, \ldots, x_k)\big) \mapsto \big(g(x_{\sigma(1)}), \ldots, g(x_{\sigma(k)})\big)$$

of the group $G = \mathfrak{S}_k \times \mathrm{PGL}(n, \mathbb{C})$.

When $k \leq n + 1$ the action of G on V is transitive and there is exactly one isomorphism class. When $k \geq n + 2$ the action is locally free (the stabilizer of any point in V is finite). In particular the set of isomorphism classes of k unordered points in $\mathbb{P}^{n-1}$ has dimension $(k - n - 1)(n - 1)$.

If $\mathcal{W}$ and $\mathcal{W}' = \varphi^* \mathcal{W}$ are two biholomorphically equivalent k-webs on $(\mathbb{C}^n, 0)$, then their tangent spaces at the origin determine two sets of k points on $\mathbb{P}^{n-1}$ which are isomorphic through $[d\varphi(0)]$, the projective automorphism determined by the projectivization of the linear map $d\varphi(0)$. It is then clear that for $k \geq n + 2$ there are many non-equivalent germs of smooth k-webs on $(\mathbb{C}^n, 0)$.

$$\S$$

It is tempting to infer from the discussion above that there is only one equivalence class of smooth $(n+1)$-web on $(\mathbb{C}^n, 0)$ using the following fallacious arguments: (a) to a $(n+1)$-web on $(\mathbb{C}^n, 0)$ one can associate $n+1$ sections of $\mathbb{P}T^*(\mathbb{C}^n, 0)$; (b) since there is only one isomorphism class of unordered $(n + 1)$ points in general position in $\mathbb{P}^{n-1}$ these sections can be sent, through a biholomorphism of $\mathbb{P}T^*(\mathbb{C}^n, 0)$, to the constant sections $[dx_1], \ldots, [dx_n], [dx_1 + \cdots + dx_n]$; (c) therefore (a) and (b) imply that every smooth $(n + 1)$-web is equivalent to the web $\mathcal{W}(dx_1, \ldots, dx_n, dx_1 + \cdots + dx_n)$.

While (a) and (b) are sound, the conclusion (c) is completely unjustified. The point is that the automorphism used in (b) is not necessarily induced by a biholomorphism $\varphi \in \mathrm{Diff}(\mathbb{C}^n, 0)$. To wit, every biholomorphism $\Phi : \mathbb{P}T^*(\mathbb{C}^n, 0) \to \mathbb{P}T^*(\mathbb{C}^n, 0)$ can be written in the form

$$\Phi(x, v) = \big(\varphi(x), [A(x) \cdot v] \big)$$

where $\varphi \in \mathrm{Diff}(\mathbb{C}^n, 0)$ and $A \in \mathrm{GL}(n, \mathcal{O}(\mathbb{C}^n, 0))$. But for very few of them, one has $[A(x) \cdot v] = [d\varphi(x) \cdot v]$ for every $(x, v) \in \mathbb{P}T^*(\mathbb{C}^n, 0)$.

It will be shown in Sect. 1.2 that not every 3-web on $(\mathbb{C}^2, 0)$ is equivalent to the parallel 3-web $\mathcal{W}(dx, dy, dx + dy)$.

1.1.3 Algebraic Webs

Given a reduced projective curve $C \subset \mathbb{P}^n$ of degree d and a hyperplane $H_0 \in \check{\mathbb{P}}^n$ intersecting C transversely, there is a natural germ of quasi-smooth d-web $\mathcal{W}_C(H_0)$ on $(\check{\mathbb{P}}^n, H_0)$ defined by the submersions $p_1, \ldots, p_d : (\check{\mathbb{P}}^n, H_0) \to C$ which describe the intersections of $H \in (\check{\mathbb{P}}^n, H_0)$ with C. Explicitly, if one writes the restriction of C to a sufficiently small neighborhood of $H_0 \subset \mathbb{P}^n$ as $C_1 \cup \cdots \cup C_d$, where the curves C_i are pairwise disjoint curves, then the functions p_i are defined as $p_i(H) = H \cap C_i$. The corresponding d-web is

$$\mathcal{W}_C(H_0) = \mathcal{W}\big(p_1, \ldots, p_d \big).$$

The d-webs of the form $\mathcal{W}_C(H_0)$ for some reduced projective curve C and some transverse hyperplane H_0 are classically called **algebraic d-webs**.

From the definition of p_i, it is clear that the inclusion

$$p_i^{-1}\big(p_i(H)\big) \subset \{H' \in \check{\mathbb{P}}^n \mid p_i(H) \in H'\}$$

holds true for every $H \in (\check{\mathbb{P}}^n, H_0)$ and every $i \in \underline{d}$. In other words the fiber of p_i through a point $H \in (\check{\mathbb{P}}^n, H_0)$ is contained in the set of hyperplanes that contain the point $p_i(H) \in C_i \subset C \subset \mathbb{P}^n$. Consequently the fibers of the submersions p_i are (pieces of) hyperplanes (Fig. 1.2).

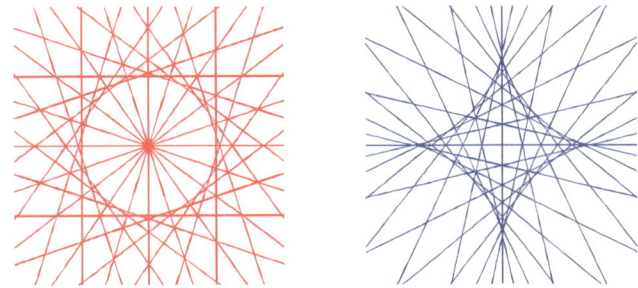

Fig. 1.2 On the *left* $\mathcal{W}_C$ is pictured for a planar reduced cubic curve C formed by a line and a conic. On the *right* $\mathcal{W}_C$ is drawn for a planar rational quartic C

It is clear from the definition of $\mathcal{W}_C(H_0)$ that when C is a reducible curve with irreducible components $C_1, \ldots, C_m$ then $\mathcal{W}_C(H_0) = \mathcal{W}_{C_1}(H_0) \boxtimes \cdots \boxtimes \mathcal{W}_{C_m}(H_0)$.

Note that the fact that C is a projective curve has not really been taken into account. If it is agreed to define **linear webs** as webs whose all leaves are (pieces of) hyperplanes, then the construction just presented establishes an equivalence between linear quasi-smooth k-webs on $(\check{\mathbb{P}}^n, H_0)$, and k germs of curves in $\mathbb{P}^n$ intersecting H_0 transversely in k distinct points (Fig. 1.3).

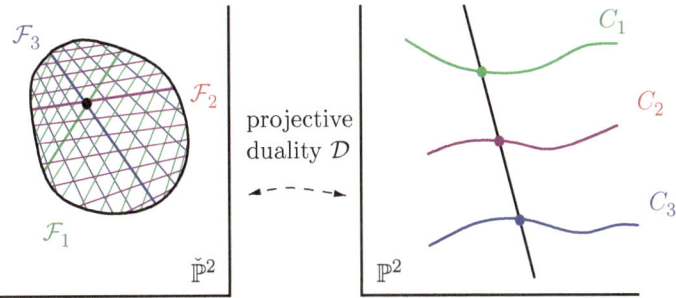

Fig. 1.3 Projective duality for a linear 3-web in dimension two

Back to the case where C is projective, if no irreducible component of C is a line, then we have the following alternative description of $\mathcal{W}_C(H_0)$. Let $\check{C}$ be the dual hypersurface of C, that is, $\check{C} \subset \check{\mathbb{P}}^n$ is the closure of the union of hyperplanes $H \in \check{\mathbb{P}}^n$ containing a tangent line of C at some smooth point $p \in C_{sm}$. Symbolically,

$$\check{C} = \overline{\bigcup_{p \in C_{sm}} \bigcup_{\substack{H \in \check{\mathbb{P}}^n \\ T_p C \subset H}} H} \ .$$

The leaves of $\mathcal{W}_C(H_0)$ through H_0 are the hyperplanes passing through it and tangent to $\check{C}$ at some point $p \in \check{C}$.

A similar interpretation holds true when C does contain lines among its irreducible components. The differences are: the dual of a line is no longer a hypersurface but a $\mathbb{P}^{n-2}$ linearly embedded in $\check{\mathbb{P}}^n$; and the leaf of a 1-web dual to a line through a point $H_0 \in \check{\mathbb{P}}^n$ is the hyperplane in $\check{\mathbb{P}}^n$ containing both H_0 and the dual $\mathbb{P}^{n-2}$.

1.2   Planar 3-Webs

This section presents one of the most emblematic results of web geometry: the characterization of hexagonal planar 3-webs through their holonomy and curvature. The exposition here follows closely [95]. The hexagonality of algebraic 3-webs is also worked out in detail, see Sect. 1.2.4. For a more detailed account, the reader can consult [53, Lecture 18].

1.2.1   Holonomy and Hexagonal Webs

Let $\mathcal{W} = \mathcal{F}_1 \boxtimes \mathcal{F}_2 \boxtimes \mathcal{F}_3$ be a germ of smooth 3-web on $(\mathbb{C}^2, 0)$. Denote by L_1, L_2, L_3 the leaves through 0 of the three foliations $\mathcal{F}_1, \mathcal{F}_2, \mathcal{F}_3$ respectively.

If $x = x_1 \in L_1$ is a point sufficiently close to the origin then, thanks to the persistence of transversal intersections under small deformations, the leaf of $\mathcal{F}_3$ through it intersects L_2 in a unique point x_2 close to the origin. Moreover the map $h_{12} : (L_1, 0) \to (L_2, 0)$ that associates the point $x_2 \in L_2$ to $x = x_1 \in L_1$ is a germ of biholomorphism.

Analogously there exists a germ of biholomorphism $h_{23} : (L_2, 0) \to (L_3, 0)$ that associates to $x_2 \in L_2$ the point $x_3 \in L_3$ defined by the intersection of L_3 with the leaf of $\mathcal{F}_1$ through $x_2 \in L_2$.

Proceeding in this way one can construct a sequence of points $x_1 \in L_1$, $x_2 \in L_2$, $x_3 \in L_3$, $x_4 \in L_1$, $x_5 \in L_2$, $x_6 \in L_3$, $x_7 \in L_1$. The function that associates to the initial point $x = x_1$ the end point x_7 is the germ of biholomorphism $h : (L_1, 0) \to (L_1, 0)$ given by the composition

$$h_{31} \circ h_{23} \circ h_{12} \circ h_{31} \circ h_{23} \circ h_{12}.$$

The reader is invited to verify the following properties of the biholomorphism h.

(a) If one does the same construction but with the roles of the foliations $\mathcal{F}_1, \mathcal{F}_2, \mathcal{F}_3$ replaced by the foliations $\mathcal{F}_{\sigma(1)}, \mathcal{F}_{\sigma(2)}, \mathcal{F}_{\sigma(3)}$—$\sigma$ being a permutation of $\{1, 2, 3\}$—then the resulting biholomorphism is conjugated to $h^{\mathrm{sign}(\sigma)}$;

(b) If $\varphi : (\mathbb{C}^2, 0) \to (\mathbb{C}^2, 0)$ is a biholomorphism and $\overline{\mathcal{W}} = \varphi^* \mathcal{W}$, then the corresponding biholomorphism $\overline{h} : (\overline{L}_1, 0) \to (\overline{L}_1, 0)$ for the leaf $\overline{L}_1 = \varphi^{-1}(L_1)$ of $\overline{\mathcal{F}}_1 = \varphi^* \mathcal{F}_1$ is equal to $\varphi_1^{-1} \circ h \circ \varphi_1$ where $\varphi_1 : (\overline{L}_1, 0) \overset{\sim}{\to} (L_1, 0)$ stands for the restriction of φ to $\overline{L}_1$.

It follows from the two properties above that the conjugacy class in $\mathrm{Diff}(\mathbb{C}, 0)$ of the group generated by h is intrinsically attached to $\mathcal{W}$. This class is by definition the **holonomy of $\mathcal{W}$ at** 0 (Fig. 1.4). It will be convenient to say that h is the holonomy of $\mathcal{W}$ at 0 instead of repeatedly referring to the conjugacy class of the group generated by it. Hopefully no confusion will arise from this abuse of terminology.

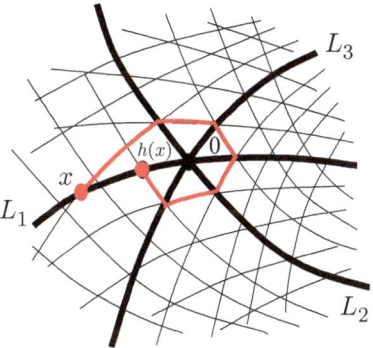

Fig. 1.4 The holonomy of a planar 3-web

To get a better grasp of the definition of the holonomy of $\mathcal{W}$ and to prepare the ground for what is to come, a family of examples parametrized by $\mathbb{C}$ is presented below in the form of a lemma.

Lemma 1.2.1. *If $\kappa \in \mathbb{C}$ is a complex number and $\mathcal{W}_\kappa = \mathcal{W}(x, y, x + y + xy$ $(x - y)(\kappa + h.o.t.))$, then the holonomy of $\mathcal{W}_\kappa$ is generated by a germ of biholomorphism $h_\kappa : (\mathbb{C}, 0) \to (\mathbb{C}, 0)$ which has as first coefficients in its series expansion*

$$h_\kappa(x) = x + 4\kappa \, x^3 + h.o.t.$$

Proof. Let $x_1 = (x, 0) \in (L_1, 0)$. To compute x_2 notice that $f_\kappa(x, y) = x + y + xy(x - y)(\kappa + h.o.t.)$ is equal to x when evaluated on both $x_1 = (x, 0)$ and $(0, x)$. In other words the leaf of $\mathcal{F}_3$, which is the foliation determined by f_κ, through x_1 cuts the leaf of $\mathcal{F}_2$, which is the foliation determined by y, in $x_2 = (0, x)$.

From the definition of x_3 it is clear that its second coordinate is equal to x. To determine its first coordinate one has to solve the implicit equation $f_\kappa(t, x) = 0$. A straightforward computation yields $t = -x - 2\kappa x^3 + h.o.t.$ and consequently $x_3 = (-x - 2\kappa x^3, x)$ up to higher order terms.

Proceeding in this way one finds

$$x_4 = \left(-x - 2\kappa x^3, 0\right) \qquad x_5 = \left(0, -x - 2\kappa x^3\right)$$

$$x_6 = \left(x + 4\kappa x^3, -x - 2kx^3\right) \qquad x_7 = \left(x + 4\kappa x^3, 0\right)$$

up to higher order terms. The details are left to the reader.   □

As far as their holonomy is concerned, the simplest smooth 3-webs on $(\mathbb{C}^2, 0)$ are the **hexagonal webs**. By definition these are the ones which can be represented in a neighborhood U of the origin by three pairwise transversal smooth foliations whose germification at any point $x \in U$ is a germ of 3-web with trivial holonomy. Using our conventions about germs, the hexagonal webs on $(\mathbb{C}^2, 0)$ are the ones with trivial holonomy at every point $x \in (\mathbb{C}^2, 0)$. The guiding example is $\mathcal{W}(x, y, x + y)$, see Fig. 1.5 below for a graphical proof of its hexagonality.

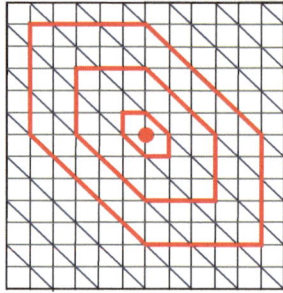

Fig. 1.5 The web $\mathcal{W}(x, y, x + y)$ is hexagonal

Beware that hexagonality is much stronger than asking the holonomy to be trivial only at the origin. It is an instructive exercise to produce an example of 3-web having trivial holonomy at zero but non-trivial holonomy at a generic $x \in (\mathbb{C}^2, 0)$.

1.2.2 Curvature for Planar 3-Webs

Suppose now that a 3-web $\mathcal{W}$ on $(\mathbb{C}^2, 0)$ is presented by its defining 1-forms, that is, $\mathcal{W} = \mathcal{W}(\omega_1, \omega_2, \omega_3)$.

There exist invertible functions $u_1, u_2, u_3 \in \mathcal{O}^*(\mathbb{C}^2, 0)$ for which

$$u_1 \omega_1 + u_2 \omega_2 + u_3 \omega_3 = 0. \tag{1.1}$$

For instance, if δ_{ij} are the holomorphic functions defined by the relation $\delta_{ij} dx \wedge dy = \omega_i \wedge \omega_j$ for $i, j = 1, 2, 3$, then

$$\delta_{23}\, \omega_1 + \delta_{31}\, \omega_2 + \delta_{12}\, \omega_3 = 0\,.$$

Although the triple $(\delta_{23}, \delta_{31}, \delta_{12})$ is not the unique solution to Eq. (1.1), any other solution differs from it by the multiplication by an invertible function. In other words, the most general solution of (1.1) is $(u_1, u_2, u_3) = u \cdot (\delta_{23}, \delta_{31}, \delta_{12})$ where $u \in \mathcal{O}^*(\mathbb{C}^2, 0)$ is arbitrary.

Lemma 1.2.2. *Let $\alpha_1, \alpha_2, \alpha_3 \in \Omega^1(\mathbb{C}^2, 0)$ be three 1-forms with pairwise wedge product nowhere zero. If $\alpha_1 + \alpha_2 + \alpha_3 = 0$, then there exists a unique 1-form η such that*

$$d\alpha_i = \eta \wedge \alpha_i \quad \text{for every } i \in \{1, 2, 3\}\,.$$

Proof. Since the ambient space has dimension two and because the 1-forms α_i are nowhere zero, there exist 1-forms γ_1, γ_2, and γ_3 satisfying

$$d\alpha_i = \gamma_i \wedge \alpha_i$$

for every $i = 1, 2, 3$. Notice that the 1-forms γ_i can be replaced by $\gamma_i + a_i \alpha_i$, with $a_i \in \mathcal{O}(\mathbb{C}^2, 0)$ arbitrary, without changing the identity above.

The difference $\gamma_1 - \gamma_2$ is again a 1-form. As such, it can be written $a_1 \alpha_1 - a_2 \alpha_2$ with $a_1, a_2 \in \mathcal{O}(\mathbb{C}^2, 0)$. Therefore

$$\gamma_1 - a_1 \alpha_1 = \gamma_2 - a_2 \alpha_2\,.$$

If $\eta = \gamma_1 - a_1 \alpha_1 = \gamma_2 - a_2 \alpha_2$, then it clearly satisfies $d\alpha_1 = \eta \wedge \alpha_1$ and $d\alpha_2 = \eta \wedge \alpha_2$. Moreover, since $\alpha_3 = -\alpha_1 - \alpha_2$, it also satisfies $d\alpha_3 = \eta \wedge \alpha_3$. This establishes the existence of η. Concerning the uniqueness, notice that two distinct solutions η and η' would verify $(\eta - \eta') \wedge \alpha_i = 0$ for $i = 1, 2, 3$. Any two of these identities are sufficient to ensure that $\eta = \eta'$. $\square$

The lemma above applied to $(\delta_{23}\omega_1, \delta_{31}\omega_2, \delta_{12}\omega_3)$ yields the existence of $\eta \in \Omega^1(\mathbb{C}^2, 0)$ such that $d(\delta_{jk}\omega_i) = \eta \wedge (\delta_{jk}\omega_i)$ for any cyclic permutation (i, j, k) of $(1, 2, 3)$. Presenting $\mathcal{W}$ through three other 1-forms—say $\omega_1' = a_1\omega_1, \omega_2' = a_2\omega_1$, and $\omega_3' = a_3\omega_3$ with $a_1, a_2, a_3 \in \mathcal{O}^*(\mathbb{C}^2, 0)$—one sees that the corresponding η' relates to η through the equation

$$\eta - \eta' = d \log(a_1 a_2 a_3)\,.$$

In particular the 1-form η depends on the presentation of $\mathcal{W}$ but in such a way that its differential does not. By definition, the 2-form $d\eta$ is the **curvature of** $\mathcal{W}$ and will be denoted by $K(\mathcal{W})$.

It seems appropriate to borrow some terminology from the nineteenth century theory of invariants and say that the 2-form $K(\mathcal{W})$ is a covariant of the web $\mathcal{W}$ since

$$K(\varphi^*\mathcal{W}) = \varphi^* K(\mathcal{W})$$

for every germ of biholomorphism $\varphi \in \text{Diff}(\mathbb{C}^2, 0)$ (the proof of this fact is straightforward and is left as an exercise to the reader).

Lemma 1.2.3. *If* $\mathcal{W} = \mathcal{W}(x, y, f)$ *where* $f \in \mathcal{O}(\mathbb{C}^2, 0)$, *then*

$$K(\mathcal{W}) = \frac{\partial^2}{\partial x \partial y}\big(\log(f_x/f_y)\big) dx \wedge dy.$$

In particular, if $\mathcal{W}_\kappa = \mathcal{W}(x, y, x + y + xy(x - y)(\kappa + h.o.t.))$, *then*

$$K(\mathcal{W}_\kappa) = \big(4\kappa + h.o.t.\big) dx \wedge dy.$$

Proof. Because $(-f_x dx) + (-f_y dy) + (df) = 0$, the 1-form η of Lemma 1.2.2 is

$$\frac{\partial}{\partial x}(\log f_y)\, dx + \frac{\partial}{\partial y}(\log f_x)\, dy$$

in this case. Hence $K(\mathcal{W})$ is as claimed.

Specializing to $\mathcal{W}_\kappa = \mathcal{W}_k(x, y, x + y + \kappa xy(x - y))$ it follows that

$$K(\mathcal{W}_\kappa) = \frac{\partial^2}{\partial x \partial y} \log\left(\frac{1 + (2xy - y^2)(\kappa + h.o.t.)}{1 - (2xy - x^2)(\kappa + h.o.t.)}\right) dx \wedge dy.$$

The second claim follows from the evaluation of the above expression at zero. □

Structure of Planar Hexagonal 3-Webs

The next result can be considered as the foundational stone of web geometry. It seems fair to say that it awakened the interest of Blaschke and his coworkers on the subject in the early 1930s.

Theorem 1.2.4. *Let* $\mathcal{W} = \mathcal{F}_1 \boxtimes \mathcal{F}_2 \boxtimes \mathcal{F}_3$ *be a smooth 3-web on* $(\mathbb{C}^2, 0)$. *The following assertions are equivalent:*

(a) *the web* $\mathcal{W}$ *is hexagonal;*
(b) *the 2-form* $K(\mathcal{W})$ *vanishes identically;*
(c) *there exist closed 1-forms* η_i *defining* $\mathcal{F}_i$, $i \in \underline{3}$ *such that* $\eta_1 + \eta_2 + \eta_3 = 0$;
(d) *the web* $\mathcal{W}$ *is equivalent to* $\mathcal{W}(x, y, x + y)$.

Beside Lemmas 1.2.1 and 1.2.3, the proof of Theorem 1.2.4 will also make use of the following:

Lemma 1.2.5. *Every germ of smooth 3-web $\mathcal{W}$ on $(\mathbb{C}^2, 0)$ is equivalent to $\mathcal{W}(x, y, f)$, where $f \in \mathcal{O}(\mathbb{C}^2, 0)$ is of the form*

$$f(x, y) = x + y + xy(x - y)(\kappa + h.o.t.)$$

for a suitable $\kappa \in \mathbb{C}$.

Proof. As already pointed out in Sect. 1.1.2 every smooth 2-web on $(\mathbb{C}^2, 0)$ is equivalent to $\mathcal{W}(x, y)$. Therefore it can be assumed that $\mathcal{W} = \mathcal{W}(x, y, g)$ with $g \in \mathcal{O}(\mathbb{C}^2, 0)$ such that $g(0) = 0$. The smoothness assumption on $\mathcal{W}$ translates into $dx \wedge dg(0) \neq 0$ and $dy \wedge dg(0) \neq 0$ or, equivalently, both $g_x(0)$ and $g_y(0)$ are non-zero complex numbers.

After pulling back $\mathcal{W}$ by $\varphi_1(x, y) = (g_x(0)x, g_y(0)y)$ one can assume that $\mathcal{W}$ still takes the form $\mathcal{W}(x, y, g)$ but now with the function g having $x + y$ as its linear term.

Let $a(t) = g(t, 0)$ and $b(t) = g(0, t)$. Clearly both a and b are germs of biholomorphisms of $(\mathbb{C}, 0)$. Let $\varphi(x, y) = (a^{-1}(x), b^{-1}(y))$ and set $h(x, y) = \varphi^* g(x, y) = g(a^{-1}(x), b^{-1}(y))$. Notice that $\varphi^* \mathcal{W}(x, y, g) = \mathcal{W}(x, y, h)$ and that h still has linear term equal to $x + y$. Moreover $h(0, t) = h(t, 0) = h(t, t)/2 = t$ up to higher order terms.

Because the germ $\alpha(t) = h(t, t)$ has derivative at zero of modulus distinct from one it follows from Poincaré Linearization Theorem [8, Chapter 3, §25.B] the existence of a germ of biholomorphism $\phi \in \mathrm{Diff}(\mathbb{C}, 0)$ conjugating α to its linear part. More succinctly,

$$\phi^{-1} \circ \alpha \circ \phi(t) = 2t.$$

After setting $\Phi(x, y) = (\phi(x), \phi(y))$ and $f = \phi^{-1} \circ h \circ \Phi$ one promptly verifies that the following equalities hold true:

$$f(t, 0) = f(0, t) = \frac{f(t, t)}{2} = t.$$

To conclude the proof it suffices to analyze the implications of the above identities to the series expansion $f(x, y) = \sum a_{ij} x^i y^j$. The reader is invited to fill in the details. □

Proof of Theorem 1.2.4

To prove that (a) implies (b), let us start by applying Lemma 1.2.5 to see that $\mathcal{W}$, at any point $p \in (\mathbb{C}^2, 0)$, is equivalent to

$$\mathcal{W}\big(x, y, x + y + xy(x - y)(\kappa + h.o.t.)\big)$$

with $\kappa \in \mathbb{C}$. Because $\mathcal{W}$ is hexagonal the holonomy at an arbitrary point $p \in (\mathbb{C}^2, 0)$ is the identity. Lemma 1.2.1 implies $\kappa = 0$. Lemma 1.2.3 in turn allows to deduce that $K(\mathcal{W})$ is also zero at an arbitrary point of $(\mathbb{C}^2, 0)$, thus proving that (a) implies (b).

Suppose now that (b) holds true and assume that $\mathcal{W} = \mathcal{W}(\omega_1, \omega_2, \omega_3)$ with the 1-forms ω_i satisfying $\omega_1 + \omega_2 + \omega_3 = 0$. Let η be the unique 1-form given by Lemma 1.2.2. Because $K(\mathcal{W}) = 0$, the 1-form η is closed. If

$$\eta_i = \exp\left(-\int \eta\right)\omega_i,$$

then

$$d\eta_i = -\eta \wedge \exp\left(-\int \eta\right)\omega_i + \exp\left(-\int \eta\right)d\omega_i = 0$$

because $d\omega_i = \eta \wedge \omega_i$. Moreover

$$\eta_1 + \eta_2 + \eta_3 = \exp\left(-\int \eta\right)(\omega_1 + \omega_2 + \omega_3) = 0.$$

This proves that (b) implies (c).

Now assuming the validity of (c), one can define

$$u_i(x) = \int_0^x \eta_i \quad \text{for } i \in \underline{3}.$$

Notice that $\varphi(x, y) = (u_1(x, y), u_2(x, y))$ is a biholomorphism, and clearly $\varphi^*\mathcal{W}(x, y, x + y) = \mathcal{W}(u_1, u_2, u_3)$ since $\eta_1 + \eta_2 = -\eta_3$. Thus $\mathcal{W}$ is equivalent to $\mathcal{W}(x, y, x + y)$.

The missing implication, (d) implies (a), has already been established in Fig. 1.5.
□

1.2.3   Germs of Hexagonal Webs on the Plane

Having Theorem 1.2.4 at hand it is natural to inquire about germs of smooth k-webs on $(\mathbb{C}^2, 0)$, $k > 3$, for which every 3-subweb is hexagonal. The k-webs having this property will also be called **hexagonal**.

The simplest examples of hexagonal k-webs are the **parallel** k-webs. These webs are the superposition of k pencils of parallel lines. They all can be written explicitly

$$\mathcal{W}(\lambda_1 x - \mu_1 y, \cdots, \lambda_k x - \mu_k y)$$

where the pairs $(\lambda_i, \mu_i) \in \mathbb{C}^2 \setminus \{0\}$ represent the slopes $[\mu_i : \lambda_i] \in \mathbb{P}(\mathbb{C}^2) = \mathbb{P}^1$ of the pencils.

More generally, if $\mathcal{L}_1, \ldots, \mathcal{L}_k$ are k pairwise distinct pencils of lines on $\mathbb{C}^2$ such that no line joining two base points passes through the origin, then $\mathcal{W} = \mathcal{L}_1 \boxtimes \cdots \boxtimes \mathcal{L}_k$, seen as a germ at the origin, is also a smooth hexagonal k-web.

A less evident family of examples was found by Bol. It consists of the germs of 5-webs defined as follows. Let $\mathcal{L}_1, \ldots, \mathcal{L}_4$ be four pencils of lines satisfying the same conditions as above, plus the extra condition that no three among the four base points of the pencils are colinear.[1] The 5-web obtained from the superposition of $\mathcal{L}_1 \boxtimes \cdots \boxtimes \mathcal{L}_4$ with the pencil of conics through the four base points is a smooth hexagonal 5-web on $(\mathbb{C}^2, 0)$. According to the relative position of the base points with respect to the origin, one obtains in this way a two-dimensional family of non-equivalent germs of smooth 5-webs on $(\mathbb{C}^2, 0)$. Any of these germs will be called **Bol's 5-web** $\mathcal{B}_5$ (Fig. 1.6). The abuse of terminology is justified by the fact that all these germs are germifications of the very same global singular 5-web (a concept to be introduced in Sect. 1.3.3) defined on $\mathbb{P}^2$.

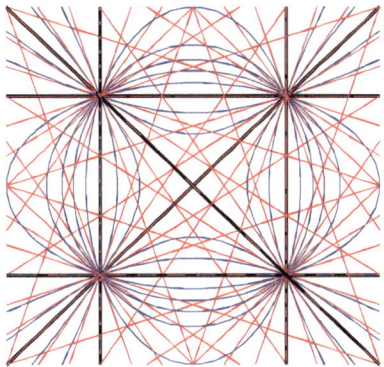

Fig. 1.6 Bol's 5-web

Anyone endeavoring to find a smooth hexagonal k-web on $(\mathbb{C}^2, 0)$ which would not be equivalent to any of the previous examples is doomed to failure. Indeed Bol proved the following.

Theorem 1.2.6. *If $\mathcal{W}$ is a smooth hexagonal k-web, $k \geq 3$, then $\mathcal{W}$ is equivalent to the superposition of k pencils of lines or $k = 5$ and $\mathcal{W}$ is equivalent to $\mathcal{B}_5$.*

A proof will not be presented here. For a recent exposition, with a fairly detailed sketch of proof, see [121].

---

[1]In other terms, the four base points of the pencils $\mathcal{L}_1, \ldots, \mathcal{L}_4$ are in 'general position' in $\mathbb{P}^2$.

1.2.4 Algebraic Planar 3-Webs Are Hexagonal

Proposition 1.2.7. *If $C \subset \mathbb{P}^2$ is a reduced cubic and $\ell_0 \subset \mathbb{P}^2$ is a line intersecting C transversely, then the 3-web $\mathcal{W}_C(\ell_0)$ is hexagonal.*

The simplest instance of the proposition above is when C is the union of three distinct concurrent lines. In this particular case it can be promptly verified that $\mathcal{W}_C(\ell_0)$ is the 3-web $\mathcal{W}(x, y, x - y)$ in a suitable affine coordinate system $[x : y : 1] \in \check{\mathbb{P}}^2$ without further ado.

In the next simplest case, C is still the union of three distinct lines but they are no longer concurrent. Then $\mathcal{W}_C(\ell_0)$ is the 3-web $\mathcal{W}(x, y, (x - 1)/(y - 1))$ in suitable affine coordinates. The most straightforward way to verify the hexagonality of this web consists in observing that the closed differential forms

$$\eta_1 = -d\log(x - 1), \quad \eta_2 = d\log(y - 1) \quad \text{and} \quad \eta_3 = d\log\big((x - 1)/(y - 1)\big)$$

define the same foliations as the submersions x, y and $(x - 1)/(y - 1)$ and satisfy $\eta_1 + \eta_2 + \eta_3 = 0$. Therefore the 3-web under scrutiny is hexagonal thanks to the equivalence between items (a) and (c) in Theorem 1.2.4.

To deal with the other cubics, one could still try to make explicit three submersions defining the web and work his way to determine a relation between closed 1-forms defining the very same foliations. Once the submersions are determined the second step is rather straightforward since the proof of Theorem 1.2.4 gives an algorithmic way to perform it. Beside having many particular cases to deal with, the lack of rational parametrizations for smooth cubics would lead to compute with Weierstrass $\wp$-functions or similar transcendental objects, adding a considerable amount of difficulty to such a task. Perhaps the most elementary way to prove the hexagonality of algebraic planar 3-webs relies on the following theorem of Chasles.

Theorem 1.2.8. *Let $X_1, X_2 \subset \mathbb{P}^2$ be two plane cubics meeting in exactly nine distinct points. If $X \subset \mathbb{P}^2$ is any cubic containing at least eight of these nine points, then it automatically contains all the nine points.*

Proof. Aiming at a contradiction, suppose that X does not contain $X_1 \cap X_2$. Let F_1, F_2 be homogenous cubic polynomials defining X_1, X_2, respectively, and G be the one defining X. Since there are nine points in the intersection of X_1 and X_2 then, according to Bezout's Theorem, the curves X_1 and X_2 must intersect transversely. In particular both curves are smooth at the intersection points. After replacing X_1 by the generic member of the pencil $\{\lambda F_1 + \mu F_2 = 0\}$ one can assume, thanks to Bertini's Theorem [63, p. 137], that X_1 is a smooth cubic. Consequently X_1 is a smooth elliptic curve.

Consider now the rational function $h : X_1 \to \mathbb{P}^1$ defined as

$$h = \left(\frac{G}{F_2}\right)\Big|_{X_1}.$$

Since X passes through eight points of $X_2 \cap X_1$ it follows that h has only one zero: the unique point of $X \cap X_1$ that does not belong to $X_1 \cap X_2$. Moreover, the transversality of X_1 and X_2 ensures that this zero is indeed a simple zero. It follows that h is an isomorphism. Since elliptic curves are not isomorphic to rational curves one arrives at a contradiction that settles the theorem.   □

The proof just presented cannot be said elementary since it makes use of Bertini's and Bezout's theorems and some basic facts of differential topology. For an elementary proof and a comprehensive account on Chasles' Theorem including its distinguished lineage and recent—rather non-elementary—developments, the reader is urged to consult [48].

Proof of Proposition 1.2.7

To deduce the hexagonality of $\mathcal{W}_C(\ell_0)$ from Chasles' Theorem, let us start by observing that the leaf of the foliation $\mathcal{F}_i$ through $\ell_0 \in \check{\mathbb{P}}^2$, denoted by L_i, corresponds to lines going through the point $p_i = p_i(\ell_0)$. To choose a point $x_1 \in L_1$ is therefore the same as choosing a line through $p_1 \in C_1 \subset \check{\mathbb{P}}^2$. If such a line is sufficiently close to ℓ_0, then it cuts C_3 in a unique point still denoted x_1. In this way the leaf L_1 of $\mathcal{F}_1$ can be identified with the curve C_3. It will also be useful to identify through the same process L_2, the leaf of $\mathcal{F}_2$ through $\ell_0 \in \check{\mathbb{P}}^2$, with C_1 and L_3 with C_2 (Fig. 1.7).

Now, following the leaf of $\mathcal{F}_3$ through $x_1 \in C_3$ until it meets L_2 corresponds to considering the line $\overline{x_1 p_2}$ and intersecting it with C_1. The intersection point $x_2 = \overline{x_1 p_2} \cap C_1 \in C_1$ corresponds to a point in L_2.

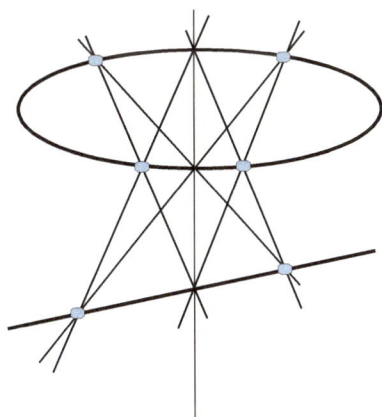

Fig. 1.7 A cubic with two irreducible components

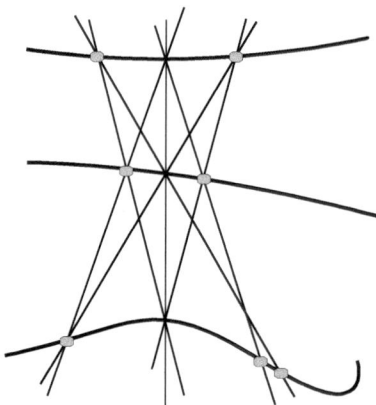

Fig. 1.8 This is not a cubic

Similarly the sequence of points $x_3, x_4, \ldots, x_7$ appearing in the definition of the holonomy of $\mathcal{W}_C(\ell_0)$ can be synthetically obtained as follows:

$$x_3 = \overline{x_2 p_3} \cap C_2 \in L_3 \simeq C_2,$$

$$x_4 = \overline{x_3 p_1} \cap C_3 \in L_1 \simeq C_3,$$

$$x_5 = \overline{x_4 p_2} \cap C_1 \in L_2 \simeq C_1,$$

$$x_6 = \overline{x_5 p_3} \cap C_2 \in L_3 \simeq C_2,$$

$$x_7 = \overline{x_6 p_1} \cap C_3 \in L_1 \simeq C_3.$$

Of course, all the identifications $L_i \simeq C_j$ above have to be understood as identifications of germs of curves (Fig. 1.8).

Notice that the line $\overline{x_1 x_2}$ is the same as $\overline{x_1 p_2}$. Therefore it contains the three points x_1, x_2, p_2. The line $\overline{x_3 x_4}$ in turn contains the points x_3, x_4, p_1 and the line $\overline{x_5 x_6}$ contains the points x_5, x_6, p_3. Thus the reduced cubic $X_1 = \overline{x_0 x_1} \cup \overline{x_2 x_3} \cup \overline{x_4 x_5}$ intersects the cubic $X_2 = C$ in exactly nine distinct points, namely $p_1, p_2, p_3, x_1, x_2, x_3, x_4, x_5, x_6$. The same reasoning shows that the reduced cubic $X = \overline{x_2 x_3} \cup \overline{x_4 x_5} \cup \overline{x_6 x_7}$ intersects $X_2 = C$ in the nine points $p_1, p_2, p_3, x_2, x_3, x_4, x_5, x_6, x_7$. Thus $X_1 \cap X_2 \cap X$ contains at least eight points. Chasles' Theorem implies that this eight is indeed a nine and consequently the points x_1 and x_7 must coincide. This is sufficient to prove that the holonomy of $\mathcal{W}_C(\ell_0)$ is the identity.   □

Later in this book, the hexagonality of the planar algebraic 3-webs will be established again using Abel's addition Theorem. Although apparently unrelated, the two approaches are intimately intertwined. Abel's addition Theorem can be read as a result about the group structure of the Jacobian of projective curves while Chasles' Theorem turns out to be equivalent to the existence of an abelian group structure for plane cubics where aligned points sum up to zero.

1.3 Singular and Global Webs

1.3.1 Germs of Singular Webs I

It is customary to say that a germ of singular holomorphic foliation is an equivalence class $[\omega]$ of germs of holomorphic 1-forms in $\Omega^1(\mathbb{C}^n, 0)$ modulo multiplication by elements of $\mathcal{O}^*(\mathbb{C}^n, 0)$ such that any representative ω is integrable (that is, such that $\omega \wedge d\omega = 0$) and has **singular set** $\text{sing}(\omega) = \{p \in (\mathbb{C}^n, 0) \, ; \, \omega(p) = 0\}$ of codimension at least two.

An analogous definition can be made for codimension one webs. A **germ of singular codimension one k-web** on $(\mathbb{C}^n, 0)$ is an equivalence class $[\omega]$ of germs of k-symmetric 1-forms, that is sections of $\text{Sym}^k \Omega^1(\mathbb{C}^n, 0)$, modulo multiplication by $\mathcal{O}^*(\mathbb{C}^n, 0)$ such that a suitable representative ω defined in a connected open neighborhood U of the origin satisfies the following conditions:

(a) the zero set of ω has codimension at least two;
(b) ω, seen as a homogeneous polynomial of degree k in the ring $\mathcal{O}(\mathbb{C}^n, 0)$ $[dx_1, \ldots, dx_n]$, is square-free;
(c) **(Brill's condition)** the germ of ω at a generic point of U is a product of k 1-forms;
(d) **(Frobenius' condition)** the germ of ω at a generic point of U is the product of k germs of integrable 1-forms.

Both conditions (c) and (d) are automatic for germs of webs on $(\mathbb{C}^2, 0)$ and non-trivial for germs on $(\mathbb{C}^n, 0)$ when $n \geq 3$. Notice also that condition (d) implies condition (c). Nevertheless the two conditions are stated independently because condition (c) is of a purely algebraic nature (it depends only on the value of ω at p) while condition (d) involves the exterior differential and therefore depends not only on the value of $\omega(p)$ but also on the local behavior of ω near p.

A germ of singular web will be called **generically smooth** if the condition below is satisfied as well:

(e) **(Generic position)** for a generic $p \in U$, any $m \leq n$ distinct germs of 1-forms $\alpha_1, \ldots, \alpha_m$ dividing the germ of ω at p satisfy

$$(\alpha_1 \wedge \cdots \wedge \alpha_m)(p) \neq 0 \, .$$

One can rephrase condition (e) by saying that for a generic $p \in U$ the germ of ω at p defines a smooth web.

§

It is interesting to compare the above definition with the following: a germ of singular codimension q foliation on $(\mathbb{C}^n, 0)$ is an equivalence class $\mathcal{F} = [\omega]$ of germs of q-forms modulo multiplication by elements of $\mathcal{O}^*(\mathbb{C}^n, 0)$ satisfying the following conditions (as above ω is a representative of $\mathcal{F}$ defined on a small open neighborhood of the origin $U \subset \mathbb{C}^n$):

(a) the zero set of ω has codimension at least two;
(c) (**Plücker's condition**) for a generic $p \in U$, the germ of ω at p is a wedge product of k linear forms $\alpha_1, \dots, \alpha_k$;
(d) (**Frobenius' condition**) for a generic $p \in U$, the germ of ω at p is the wedge product of k germs of 1-forms $\alpha_1, \dots, \alpha_k$ and each one of them satisfies $d\alpha_i \wedge \omega = 0$.

Notice that the absence of condition (b) is due to the antisymmetric character of $\Omega^q(\mathbb{C}^n, 0)$. Although apparently similar, conditions (c), (d) for codimension q foliations and k-webs have rather distinct features.

It is a classical result of Plücker that the q-form $\omega(p)$ satisfies condition (c) (foliation version) if and only if[2]

$$\left(i_v \omega(p)\right) \wedge \omega(p) = 0 \quad \text{for every} \quad v \in \bigwedge^{q-1} T_p \mathbb{C}^n .$$

Moreover, varying $v \in \bigwedge^{q-1} T_p \mathbb{C}^n$, the above formulas are the well-known Plücker quadrics and generate the homogeneous ideal defining the locus of completely decomposable q-forms in $\bigwedge^q T_p^* \mathbb{C}^n$, see, for instance, [63, pp. 209–211] or [55, Chapter 3,Theorem 1.5].

Less well known are Brill's equations describing the locus of completely decomposable k-symmetric 1-forms, see [55, Chapter 4, Section 2] for a modern exposition. They differ from Plücker equations in a number of ways: they cannot be so easily described since their definition depends on some concepts of representation theory; they are not quadratic equations, for k-symmetric 1-forms they are of degree $k + 1$; the ideal generated by Brill's equations is not reduced in general; already for 3-symmetric 1-forms on $\mathbb{C}^4$ Brill's equations do not generate the ideal of the locus of completely decomposable forms, see [39, Proposition 2.5] for instance.

For condition (d) the situation is even worse. While for alternate q-forms the integrability condition can be written

$$(i_v d\omega) \wedge \omega = 0 \quad \text{for every} \quad v \in \bigwedge^{q-1} T_p \mathbb{C}^n ,$$

(see [42, Propositions 1.2.1 and 1.2.2]), the integrability condition for k-symmetric 1-forms has not been treated yet in the literature.

1.3.2 Germs of Singular Webs II

There is an alternative definition for germs of singular webs that is in a certain sense more geometric. The idea is to consider the (germ of) web as a meromorphic section

---

[2]Here, and throughout, i_v denotes the interior product.

of the projectivization of the cotangent bundle. This is a classical point of view in the theory of differential equations which has been recently explored in web geometry by Cavalier–Lehmann, see [26]. Beware that the terminology here adopted does not always coincide with the one used in [26].

The Contact Distribution

Let $\mathbb{P} = \mathbb{P}T^*(\mathbb{C}^n, 0)$ be the projectivization of the cotangent bundle of $(\mathbb{C}^n, 0)$ and $\pi : \mathbb{P} \to (\mathbb{C}^n, 0)$ the natural projection.[3] On $\mathbb{P}$ there is a canonical codimension one distribution, the so-called **contact distribution** $\mathcal{D}$. Its description in terms of a system of coordinates $x_1, \ldots, x_n$ of $(\mathbb{C}^n, 0)$ goes as follows: if $y_i = \partial_i$ are interpreted as coordinates[4] of the total space of $T^*(\mathbb{C}^n, 0)$, then the lift of the contact distribution from $\mathbb{P}$ to $T^*(\mathbb{C}^n, 0)$ is the kernel of the 1-form

$$\alpha = \sum_{i=1}^{n} y_i dx_i. \tag{1.2}$$

The usual way to define $\mathcal{D}$ in more intrinsic terms goes as follows. Recall that the tautological line-bundle $\mathcal{O}_{\mathbb{P}}(-1)$ is the rank one subbundle of $\pi^*T^*(\mathbb{C}^n, 0)$ determined over a point $p = (x, [y]) \in \mathbb{P}$ by the direction parametrized by it. Its dual $\mathcal{O}_{\mathbb{P}}(1)$ is therefore a quotient of $\pi^*T(\mathbb{C}^n, 0)$. The distribution on $\mathbb{P}$ induced by the kernel of the composition

$$T\mathbb{P} \xrightarrow{d\pi} \pi^*T(\mathbb{C}^n, 0) \longrightarrow \mathcal{O}_{\mathbb{P}}(1)$$

is nothing else than the contact distribution $\mathcal{D}$. Notice that the composition is given by the interior product of local sections of $T\mathbb{P}$ with a twisted 1-form $\alpha \in H^0(\mathbb{P}, \Omega_{\mathbb{P}}^1 \otimes \mathcal{O}_{\mathbb{P}}(1))$, which in local coordinates coincides with the 1-form (1.2). This 1-form is the so-called **contact form** of $\mathbb{P}$.

Webs as Closures of Meromorphic Multi-sections

Let now $W \subset \mathbb{P}$ be a subvariety not necessarily irreducible but of pure dimension n. Suppose also that W satisfies the following conditions:

(a) the image under π of every irreducible component of W has dimension n;

---

[3]The convention adopted in this text is that over a point p the fiber $\pi^{-1}(p)$ is the space of one-dimensional subspaces of $T_p^*(\mathbb{C}^n, 0)$. Beware that some authors consider $\pi^{-1}(p)$ as the space of one-dimensional quotients of $T_p^*(\mathbb{C}^n, 0)$.

[4]In case of confusion, notice that the coordinate functions on a vector space E can be chosen to be elements of E^*, that is, linear forms on E.

(b) the generic fiber of π intersects W in k distinct smooth points at which the differential $d\pi_{|W} : T_p W \to T_{\pi(p)}(\mathbb{C}^n, 0)$ is surjective;
(c) the restriction of the contact form α to the smooth part of every irreducible component of W is integrable.

One can then define a germ of web as a subvariety W of $\mathbb{P}$ as above. This definition is equivalent to the one laid down in Sect. 1.3.1. Indeed given a singular k-web $[\omega]$ in the sense of Sect. 1.3.1 one can consider the closure of its "graph" in $\mathbb{P}$. More precisely, over a generic point $p \in (\mathbb{C}^n, 0)$ the "graph" of $[\omega(p)]$ is formed by the points in $\mathbb{P}T_p^*(\mathbb{C}^n, 0)$ corresponding to the factors of $\omega(p)$. In this way one defines a locally closed subvariety of $\mathbb{P}$ whose closure satisfies conditions (a), (b), and (c) above.

Reciprocally the restriction of the contact form α to a subvariety $W \subset \mathbb{P}$ satisfying (a), (b), and (c) above induces a codimension one foliation $\mathcal{F}$ on the smooth part of W. Moreover, over regular values of π the direct image of $\mathcal{F}$ by π can be identified with the superposition of k foliations. Since the symmetric product of the k distinct 1-forms defining these foliations is invariant under the monodromy of π, one ends up with a germ of section of $\mathrm{Sym}^k \Omega^1(\mathbb{C}^n, 0)$ inducing $\pi_* \mathcal{F}$. After cleaning up eventual codimension one components of the zero set, one obtains a k-symmetric 1-form ω satisfying the conditions (a), (b), (c), and (d) of Sect. 1.3.1.

1.3.3   Global Webs

Although this book is ultimately interested in the classification of germs of smooth webs of maximal rank, a concept to be introduced in Chap. 2, most of the relevant examples are globally defined on projective manifolds. It is therefore natural to lay down the definitions of a global web and the related concepts.

A **global k-web** $\mathcal{W}$ on a manifold X is given by an open covering $\mathcal{U} = \{U_i\}$ of X and k-symmetric 1-forms $\omega_i \in \mathrm{Sym}^k \Omega_X^1(U_i)$ subject to the following conditions:

1. for each non-empty intersection $U_i \cap U_j$ of elements of $\mathcal{U}$ there exists a non-vanishing holomorphic function $g_{ij} \in \mathcal{O}^*(U_i \cap U_j)$ such that $\omega_i = g_{ij} \omega_j$;
2. for every $U_i \in \mathcal{U}$ and every $x \in U_i$, the germification of ω_i at x satisfies the conditions (a), (b), (c), and (d) of Sect. 1.3.1. In other words, the germ of ω_i at x is a representative of a germ of a singular web.

The transition functions g_{ij} determine a line-bundle $\mathcal{N}$ over X and the k-symmetric 1-forms $\{\omega_i\}$ patch together to form a global section of $\mathrm{Sym}^k \Omega_X^1 \otimes \mathcal{N}$, that is, $\omega = \{\omega_i\}$ can be interpreted as an element of $H^0(X, \mathrm{Sym}^k \Omega_X^1 \otimes \mathcal{N})$. The line-bundle $\mathcal{N}$ will be called the **normal bundle** of $\mathcal{W}$. Two global sections $\omega, \omega' \in H^0(X, \mathrm{Sym}^k \Omega_X^1 \otimes \mathcal{N})$ determine the same web if and only if they differ by the multiplication by an element $g \in H^0(X, \mathcal{O}_X^*)$.

If X is compact, or more generally if the only global sections of $\mathcal{O}_X^*$ are the constants, then a global k-web is nothing more than an element of

$\mathbb{P} H^0(X, \mathrm{Sym}^k \Omega^1_X \otimes \mathcal{N})$, for a suitable line-bundle $\mathcal{N} \in \mathrm{Pic}(X)$, with germification of any representative at any point of X satisfying conditions (a), (b), (c), and (d) of Sect. 1.3.1.

When X is a variety such that every line-bundle over it has non-zero meromorphic sections, one can alternatively define global k-webs as equivalence classes $[\omega]$ of meromorphic k-symmetric 1-forms modulo multiplication by meromorphic functions such that the germification of any representative ω at a generic point of X satisfies the very same conditions referred to above. The transition to the previous definition is made by observing that a meromorphic k-symmetric 1-form ω can be interpreted as a global holomorphic section of $\mathrm{Sym}^k \Omega^1_X \otimes \mathcal{O}_X((\omega)_\infty - (\omega)_0)$ where $(\omega)_0$, respectively $(\omega)_\infty$, stands for the zero divisor, respectively the polar divisor, of ω.

A global k-web $\mathcal{W} \in \mathbb{P} H^0(X, \mathrm{Sym}^k \Omega^1_X \otimes \mathcal{N})$ is **decomposable** if there are global webs $\mathcal{W}_1, \mathcal{W}_2$ on X sharing no common subwebs such that $\mathcal{W}$ is the superposition of $\mathcal{W}_1$ and $\mathcal{W}_2$, that is $\mathcal{W} = \mathcal{W}_1 \boxtimes \mathcal{W}_2$. A k-web $\mathcal{W}$ will be called **completely decomposable** if one can write $\mathcal{W} = \mathcal{F}_1 \boxtimes \cdots \boxtimes \mathcal{F}_k$ for k global foliations $\mathcal{F}_1, \dots, \mathcal{F}_k$ on X. Remark that the restriction of a web on a sufficiently small neighborhood of a generic $x \in X$ is completely decomposable.

Monodromy

Thanks to condition (b) of Sect. 1.3.1, the germ $\mathcal{W}_x$ of a global k-web $\mathcal{W}$ at a generic point $x \in X$ is completely decomposable. Moreover the set of points $x \in X$ where $\mathcal{W}_x$ is not completely decomposable is a closed analytic subset of X. If U is the complement of this subset, then for arbitrary $x_0 \in U$ it is possible to write $\mathcal{W}_{x_0}$ as $\mathcal{F}_1 \boxtimes \cdots \boxtimes \mathcal{F}_k$. Notice that $\mathcal{W}$ does not have to be quasi-smooth at $x_0 \in U$. It may happen that $T_{x_0} \mathcal{F}_i = T_{x_0} \mathcal{F}_j$ for some $i \neq j$.

Analytic continuation of this decomposition along paths γ contained in U determines an anti-homomorphism[5]

$$\rho_\mathcal{W} : \pi_1(U, x_0) \longrightarrow \mathfrak{S}_k$$

from the fundamental group of U to the permutation group on k letters $\mathfrak{S}_k$. Because distinct choices of base points yield conjugated anti-homomorphisms, it is harmless to identify all these anti-homomorphisms and call them the monodromy (anti)-representation of $\mathcal{W}$. The image of $\rho_\mathcal{W} \subset \mathfrak{S}_k$ is, by definition, the **monodromy group** of $\mathcal{W}$.

The reader is invited to verify the validity of the following proposition.

---

[5]As usual the fundamental group acts on the right and thus $\rho_\mathcal{W}$ is not a homomorphism but instead an anti-homomorphism, that is one has $\rho_\mathcal{W}(\gamma_1 \cdot \gamma_2) = \rho_\mathcal{W}(\gamma_2) \cdot \rho_\mathcal{W}(\gamma_1)$ for arbitrary $\gamma_1, \gamma_2 \in \pi_1(U, x_0)$.

Proposition 1.3.1. *The following assertions hold true.*

(a) *If $\mathcal{W}$ is not completely decomposable, then there exists $\gamma \in \pi_1(U, x_0)$ such that $\rho_{\mathcal{W}}(\gamma)$ is a non-trivial permutation;*

(b) *Every irreducible component of the complement of U has codimension one.*

Proposition 1.3.1 makes clear that $\rho_{\mathcal{W}}$ measures the obstruction for $\mathcal{W}$ to be completely decomposable.

Alternatively one can also define a global k-web on X as a closed subvariety $W \subset \mathbb{P}TX$ satisfying the natural global analogues of conditions (a), (b), and (c) of Sect. 1.3.2. In this alternative definition, the monodromy is nothing more than the usual monodromy of the projection $\pi|_W \colon W \to X$.

1.3.4 Discriminant

The discriminant locus of a k-web $\mathcal{W}$ on a complex manifold X is formed by the set of points where the germ of $\mathcal{W}$ is not quasi-smooth. Thinking $\mathcal{W}$ as a subvariety $W \subset \mathbb{P}(TX)$, the discriminant is precisely the image under the natural projection $\pi|_W \colon W \to X$ of the union of the singular points of W with the critical set of the restriction of $\pi|_W$ to the smooth locus of W.

From its very definition it is clear that the discriminant locus is a closed analytic subset whose complement V is contained in the subset U used in the definition of the monodromy representation. Therefore the monodromy representation can be seen as an anti-homomorphism from $\pi_1(V)$ to $\mathfrak{S}_k$.

As for webs $\mathcal{W}$ on surfaces, their discriminants have simple expressions inherited from the classical invariant theory of binary forms.

The Resultant and Tangencies Between Webs on Surfaces

Recall that for two homogeneous polynomials in two variables, also known as binary forms,

$$P = \sum_{i=0}^{m} p_i x^i y^{m-i} \quad \text{and} \quad Q = \sum_{i=0}^{n} q_i x^i y^{n-i}$$

the resultant $R[P, Q]$ of P and Q is given by the determinant of the Sylvester matrix

$$\begin{bmatrix} p_m & \cdots & \cdots & p_0 & & \\ & \ddots & & & \ddots & \\ & & p_m & \cdots & \cdots & p_0 \\ q_n & \cdots & q_0 & & & \\ & \ddots & & \ddots & & \\ & & q_n & \cdots & q_0 & \end{bmatrix} .$$

This is the $(m + n) \times (m + n)$-matrix formed from the coefficients of P and Q as schematically presented above with $n = \deg(Q)$ rows built from the coefficients of P and $m = \deg(P)$ rows coming from the coefficients of Q.

If $\lambda, \mu \in \mathbb{C}^*$ and if $g \in GL_2(\mathbb{C})$ is a linear automorphism of $\mathbb{C}^2$, then the resultant obeys the transformation rules

$$
\begin{aligned}
R[\lambda P, \mu Q] &= \lambda^{\deg(Q)} \mu^{\deg(P)} R[P, Q], \\
R[g^* P, g^* Q] &= \det(g)^{\deg(P) \cdot \deg(Q)} R[P, Q].
\end{aligned}
\tag{1.3}
$$

Moreover $R[P, Q]$ vanishes if and only if P and Q share a common root.

If, for $\ell = 1, 2$, $\mathcal{W}_\ell = [\omega_\ell] \in \mathbb{P}H^0(S, \mathrm{Sym}^{k_\ell} \Omega_S^1 \otimes \mathcal{N}_\ell)$ is a k_ℓ-web on a surface S, then for any i the local defining symmetric 1-forms $\omega_{\ell,i} \in \mathrm{Sym}^{k_\ell} \Omega_S^1(U_i)$ can be interpreted as binary forms in the variables dx, dy with coefficients in $\mathcal{O}_S(U_i)$. The resultant $R[\omega_{1,i}, \omega_{2,i}]$ is then an element of $\mathcal{O}_S(U_i)$ with zero locus coinciding with the tangencies between $\mathcal{W}_1|_{U_i}$ and $\mathcal{W}_2|_{U_i}$. The transformation rules (1.3) imply that the collection $\{R[\omega_{1,i}, \omega_{2,i}]\}$ patch together to form a global holomorphic section of the line-bundle $K_S^{\otimes k_1 \cdot k_2} \otimes \mathcal{N}_1^{\otimes k_2} \otimes \mathcal{N}_2^{\otimes k_1}$. This section is different from the zero section if and only if $\mathcal{W}_1$ and $\mathcal{W}_2$ do not share a common subweb since the resultant vanishes only when its parameters share common roots.

If $\mathrm{tang}(\mathcal{W}_1, \mathcal{W}_2)$ is defined as the divisor locally given by the resultant of the defining k_ℓ-symmetric 1-forms, then the preceding discussion can be summarized in the following proposition.

Proposition 1.3.2. *Let $\mathcal{W}_1$ be a k_1-web and $\mathcal{W}_2$ be a k_2-web with respective normal bundles $\mathcal{N}_1$ and $\mathcal{N}_2$, both defined on the same surface S. If they do not share a common subweb, then the identity*

$$
\mathcal{O}_S\big(\mathrm{tang}(\mathcal{W}_1, \mathcal{W}_2)\big) = K_S^{\otimes k_1 \cdot k_2} \otimes \mathcal{N}_1^{\otimes k_2} \otimes \mathcal{N}_2^{\otimes k_1}
$$

holds true in the Picard group of S.

The Discriminant of Webs on a Surface

By definition, the **discriminant** $\Delta(P)$ of a binary form $P = \sum_{i=0}^{n} p_i x^i y^{n-i}$ of degree n is

$$
\Delta(P) = \frac{R[P, \partial_x P]}{n^n p_n} = \frac{R[P, \partial_y P]}{n^n p_0}.
$$

Notice that $\Delta(P)$ vanishes if and only if both P and $\partial_x P$ share a common root, that is, if and only if P has a root with multiplicity greater than one.

The discriminant obeys rules analogous to the ones satisfied by the resultant. Namely

$$\Delta(\lambda P) = \lambda^{2(\deg(P)-1)}\Delta(P),$$

$$\Delta(g^*P) = \det(g)^{\deg(P)(\deg(P)-1)}\Delta(P).$$

If $\mathcal{W}$ is a k-web, $k \geq 2$, on a surface S, then the discriminant divisor $\Delta(\mathcal{W})$ of $\mathcal{W}$ is, by definition, the divisor locally defined by $\Delta(\omega_i)$ where as before $\omega_i \in \mathrm{Sym}^k\Omega^1_S(U_i)$ locally defines $\mathcal{W}$. Notice that the support of the discriminant divisor coincides with the discriminant set of $\mathcal{W}$ previously defined.

The discussion about the tangency of two webs adapts verbatim to yield the proposition below.

Proposition 1.3.3. *If $\mathcal{W}$ is a k-web with normal-bundle $\mathcal{N}$ defined on a surface S, then*

$$\mathcal{O}_S\big(\Delta(\mathcal{W})\big) = K_S^{\otimes k(k-1)} \otimes \mathcal{N}^{\otimes 2(k-1)}.$$

Discriminants of Real Webs

Due to obvious technical constraints, all the pictures of planar webs presented in this book are drawn over the real plane. In particular the webs portrayed ought to be defined by real analytic k-symmetric 1-forms ω on some open subset U of $\mathbb{R}^2$. Most of the time these 1-forms will be polynomial 1-forms and hence globally defined on $\mathbb{R}^2$.

The sign of the discriminant of ω at a given point $p \in U$ gives clues about the number of real leaves of $\mathcal{W} = [\omega]$ through p. For a 2-web $\mathcal{W}$ induced by $\omega = adx^2 + bdxdy + cdy^2$ the sign of $\Delta = \Delta(\omega) = b^2 - 4ac$ tells everything there is to know about the number of real leaves: when $\Delta(p) > 0$ there are two real leaves through p, and when $\Delta(p) < 0$ there is no real leaf through p.

For 3-webs the situation is as good as for 2-webs. According to whether the sign of Δ is positive or negative at a given point p the 3-web has one or three real leaves through p.

For k-webs with $k \geq 4$ the sign of Δ at p does not determine the number of real leaves of $\mathcal{W}$ through it but does tell that, see [97],

- when k is odd, the number of real leaves through p is congruent to 1 or 3 modulo 4 according as $\Delta(p) > 0$ or $\Delta(p) < 0$;
- when k is even, the number of real leaves through p is congruent to 0 or 2 modulo 4 according as $\Delta(p) < 0$ or $\Delta(p) > 0$.

It is tempting to claim that a planar k-web $\mathcal{W}$ on $\mathbb{C}^2$ defined by a real k-symmetric 1-form with only one leaf through each point of a given domain $U \subset \mathbb{R}^2$ is nothing more than an analytic foliation on U. Although trivially true if the discriminant of $\mathcal{W}$ does not intersect U this claim is far from being true in general.

Perhaps the simplest example comes from a variation on the classical Tait–Kneser Theorem presented in [57, 126]. Let $f \in \mathbb{R}[x]$ be a polynomial in one real variable of degree k. For fixed $n < k$ and $t \in \mathbb{R}$, let us define the n-th osculating polynomial g_t of f as the polynomial of degree at most n whose graph osculates the graph of f at $(t, f(t))$ up to order n. From its definition it follows that $g_t(x)$ is nothing else than the truncation of the Taylor series of f centered at t at order $n + 1$, that is

$$g_t(x) = \sum_{i=0}^{n} \frac{f^{(i)}(t)}{i!}(x - t)^i .$$

Notice that for a fixed $t \in \mathbb{C}$ for which $f^{(n)}(t) \neq 0$ the graph $G_t = \{y = g_t(x)\}$ of g_t is an irreducible plane curve of degree n. Moreover varying $t \in \mathbb{C}$ one obtains a family of degree n curves which corresponds to a degree k curve Γ_f on the space of degree n curves. The degree n curves through a generic point $p \in \mathbb{C}^2$ determine a hyperplane H in the space of degree n curves. Because H intersects Γ_f in k points, through a generic $p \in \mathbb{C}^2$ pass k distinct curves of the family $\{G_t\}$. Therefore this family of curves determines a k-web $\mathcal{W}_f$ on $\mathbb{C}^2$.

To obtain a polynomial k-symmetric 1-form defining $\mathcal{W}_f$ it suffices to eliminate t from the pair of equations

$$\begin{aligned}
y - g_t(x) &= 0, \\
dy - \partial_x g_t(x)dx &= 0.
\end{aligned}$$

Such a task can be performed by considering the resultant of $y - g_t(x)$ and $dy - \partial_x g_t(x)dx$ seen as a degree k polynomial in the variable t with coefficients in $\mathbb{C}[x, y, dx, dy]$.

To investigate the real trace of the k-web $\mathcal{W}_f$ the following variant of the classical Tait–Kneser Theorem [57, 126] will be useful.

Theorem 1.3.4. *If n is even and if $f^{(n+1)}(t) \neq 0$ for every real number t in a non-empty interval $]a, b[$, then the curves G_a and G_b do not intersect in $\mathbb{R}^2$.*

Proof. On the one hand, if $a < b$ are real numbers for which G_a and G_b intersect in $\mathbb{R}^2$, then there exists a real number $x_0 \in \mathbb{R}$ such that $g_a(x_0) - g_b(x_0) = 0$.

On the other hand, the fundamental theorem of calculus implies

$$\begin{aligned}
g_a(x_0) - g_b(x_0) &= \int_a^b \frac{\partial g_t}{\partial t}(x_0)dt \\
&= \int_a^b \left(\sum_{i=0}^{n} \frac{f^{(i+1)}(t)}{i!}(x_0 - t)^i - \sum_{i=1}^{n} \frac{f^{(i)}(t)}{(i-1)!}(x_0 - t)^{i-1} \right) dt \\
&= \int_a^b \frac{f^{(n+1)}(t)}{n!}(x_0 - t)^n dt \neq 0 .
\end{aligned}$$

This contradiction concludes the proof. $\square$

If $f \in \mathbb{R}[x]$ is a function of odd degree k, then the real trace of the k-web $\mathcal{W}_f$ defined by the graph of the $(k-1)$-th osculating functions of f is a continuous foliation in a neighborhood of Γ_f, i.e. the real graph of f, which is not differentiable since every point of Γ_f is tangent to some leaf of the foliation without being a leaf itself. To summarize: the real trace of a holomorphic (or even polynomial) web can be a non-differentiable, although continuous, foliation (Fig. 1.9).

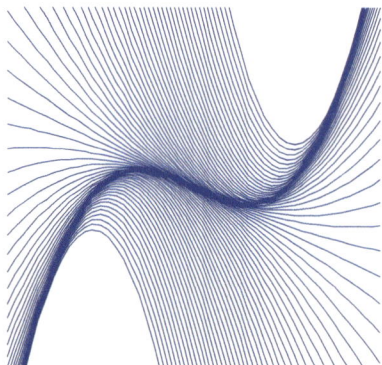

Fig. 1.9 The real trace of the 3-web $\mathcal{W}_f$ for $f = x(2x-1)(2x+1)$

1.4 Examples

1.4.1 Global Webs on Projective Spaces

Let $\mathcal{W} = [\omega] \in \mathbb{P}H^0(\mathbb{P}^n, \mathrm{Sym}^k \Omega^1_{\mathbb{P}^n} \otimes \mathcal{N})$ be a k-web on $\mathbb{P}^n$. The **degree of** $\mathcal{W}$ is defined as the number of tangencies, counted with multiplicities, of $\mathcal{W}$ with a line not everywhere tangent to $\mathcal{W}$. More precisely, if $i : \mathbb{P}^1 \to \mathbb{P}^n$ is a linear embedding of $\mathbb{P}^1$ into $\mathbb{P}^n$, then the points of tangency of the image line $i(\mathbb{P}^1)$ with $\mathcal{W}$ correspond to the zeros of $[i^*\omega] \in \mathbb{P}H^0(\mathbb{P}^1, \mathrm{Sym}^k \Omega^1_{\mathbb{P}^1} \otimes i^*\mathcal{N})$. Notice that $i^*\omega$ vanishes identically if and only if the image of i is everywhere tangent to $\mathcal{W}$.

Recall that every line-bundle $\mathcal{L}$ on $\mathbb{P}^n$ is an integral multiple of $\mathcal{O}_{\mathbb{P}^n}(1)$ and consequently one can write $\mathcal{L} = \mathcal{O}_{\mathbb{P}^n}(\deg(\mathcal{L}))$. Because the embedding i is linear, the identity $i^*\mathcal{O}_{\mathbb{P}^n}(1) = \mathcal{O}_{\mathbb{P}^1}(1)$ holds true. Putting these two facts together with the identity $\mathrm{Sym}^k \Omega^1_{\mathbb{P}^1} = \mathcal{O}_{\mathbb{P}^1}(-2k)$ yields

$$\deg(\mathcal{W}) = \deg(\mathcal{N}) - 2k .$$

Characteristic Numbers of Projective Webs

Let $X \subset \mathbb{P}^n$ be an irreducible projective subvariety. The projectivized conormal variety of X, the **conormal variety** of X for short, is the unique closed subvariety $\mathrm{Con}(X)$ of $\mathbb{P}T^*\mathbb{P}^n$ satisfying:

1. $\pi(\mathrm{Con}(X)) = X$, where $\pi : \mathbb{P}T^*\mathbb{P}^n \to \mathbb{P}^n$ is the natural projection;
2. the fiber $\pi^{-1}(x) \cap \mathrm{Con}(X)$ over any smooth point x of X is $\mathbb{P}T_x^*X \subset \mathbb{P}T^*\mathbb{P}^n$.

The conormal variety of $X \subset \mathbb{P}^n$ can succinctly be defined as

$$\mathrm{Con}(X) = \overline{\mathbb{P}N^*X_{sm}},$$

with X_{sm} denoting the smooth part of X and N^*X_{sm} its conormal bundle.

For example, the conormal of a point $x \in \mathbb{P}^n$ is all the fiber $\pi^{-1}(x) = \mathbb{P}T_x^*\mathbb{P}^n$. More generally the conormal variety of a linearly embedded $\mathbb{P}^i \subset \mathbb{P}^n$ is a trivial $\mathbb{P}^{n-i-1}$ bundle over $\mathbb{P}^i$.

If $W \subset \mathbb{P}T^*\mathbb{P}^n$ is the natural lift of $\mathcal{W}$ (cf. Sect. 1.3.2 above), then the **characteristic numbers** of $\mathcal{W}$ on $\mathbb{P}^n$ are, by definition, the n integers

$$d_i(\mathcal{W}) = W \cdot \mathrm{Con}(\mathbb{P}^i), \qquad i = 0, \ldots, n-1,$$

where $A \cdot B$ stands for the intersection product of two algebraic cycles A and B on $\mathbb{P}T^*\mathbb{P}^n$.

Notice that $d_0(\mathcal{W})$ counts the number of leaves of $\mathcal{W}$ through a generic point of $\mathbb{P}^n$, that is $\mathcal{W}$ is a $d_0(\mathcal{W})$-web. The integer $d_1(\mathcal{W})$ counts the number of points over a generic line ℓ where the web has a leaf whose tangent space contains ℓ. Therefore $d_1(\mathcal{W})$ is nothing else than the degree of $\mathcal{W}$ defined previously.

1.4.2 Algebraic Webs Revisited

It seems fair to say that the simplest k-webs on projective spaces are the ones of degree zero. Perhaps the best way to describe them is through projective duality.

Let $\check{\mathbb{P}}^n$ denote the projective space parametrizing hyperplanes in $\mathbb{P}^n$ and let $\mathcal{I} \subset \mathbb{P}^n \times \check{\mathbb{P}}^n$ be the incidence variety, that is

$$\mathcal{I} = \big\{ (p, H) \in \mathbb{P}^n \times \check{\mathbb{P}}^n \,\big|\, p \in H \big\}.$$

The natural projections from $\mathcal{I}$ to $\mathbb{P}^n$ and $\check{\mathbb{P}}^n$ will be, respectively, denoted by π and $\check{\pi}$.

Proposition 1.4.1. *The incidence variety $\mathcal{I}$ is naturally isomorphic to $\mathbb{P}T^*\mathbb{P}^n$ and also to $\mathbb{P}T^*\check{\mathbb{P}}^n$. Moreover, under these isomorphisms the natural projections π and $\check{\pi}$ from $\mathcal{I}$ to $\mathbb{P}^n$ and $\check{\mathbb{P}}^n$ coincide with the projections from $\mathbb{P}T^*\mathbb{P}^n$ to $\mathbb{P}^n$ and from $\mathbb{P}T^*\check{\mathbb{P}}^n$ to $\check{\mathbb{P}}^n$, respectively.*

Proof. If one identifies $\mathbb{P}^n \times \check{\mathbb{P}}^n$ with $\mathbb{P}(V) \times \mathbb{P}(V^*)$ where V is a vector space V of dimension $n + 1$, then the incidence variety can be identified with the projectivization of the locus defined on $V \times V^*$ through the vanishing of the natural pairing. Combining this with the natural isomorphism between V and V^{**} the proposition follows. For details see [55, Chap. 1,3.A]   □

Using this identification of $\mathcal{I}$ with $\mathbb{P}T^*\check{\mathbb{P}}^n$ one defines for every projective curve $C \subset \mathbb{P}^n$ its dual web $\mathcal{W}_C$ as the one defined by the variety $\pi^{-1}(C) \subset \mathbb{P}T^*\check{\mathbb{P}}^n$ seen as a multi-section of $\check{\pi} : \mathbb{P}T^*\check{\mathbb{P}}^n \to \check{\mathbb{P}}^n$. One verifies at once that the germification of this global web at a generic point $H_0 \in \check{\mathbb{P}}^n$ coincides with the germ of web $\mathcal{W}_C(H_0)$ defined in Sect. 1.1.3.

Proposition 1.4.2. *If $C \subset \mathbb{P}^n$ is a projective curve of degree k, then $\mathcal{W}_C$ is a k-web of degree zero on $\check{\mathbb{P}}^n$. Reciprocally, if $\mathcal{W}$ is a k-web of degree zero on $\check{\mathbb{P}}^n$, then there exists $C \subset \mathbb{P}^n$, a projective curve of degree k, such that $\mathcal{W} = \mathcal{W}_C$.*

Proof. If $C \subset \mathbb{P}^n$ is a projective curve, then all the leaves of $\mathcal{W}_C$ are hyperplanes. Therefore a line tangent to $\mathcal{W}_C$ at a quasi-smooth point is automatically contained in the leaf passing through that point. This is sufficient to prove that $\mathcal{W}_C$ has degree 0. Alternatively one can compute directly $d_1(\mathcal{W}_C) = \pi^{-1}(C) \cdot \mathrm{Con}(\check{\mathbb{P}}^1)$ for a general line $\check{\mathbb{P}}^1$ in $\check{\mathbb{P}}^n$. The conormal variety of a line $\check{\mathbb{P}}^1$ in $\check{\mathbb{P}}^n$ coincides with the conormal variety of a general $\mathbb{P}^{n-2}$ in $\mathbb{P}^n$. Since C does not intersect a general codimension two subspace of $\mathbb{P}^n$ we deduce that $d_1(\mathcal{W}_C) = 0$.

The proof of the reciprocal is similar and the reader is invited to work it out.   □

The Discriminant of $\mathcal{W}_C$

If C is a smooth projective curve, then the discriminant of $\mathcal{W}_C$ is nothing else than the set of hyperplanes tangent to C at some point. Succinctly, when C is smooth:

$$\Delta(\mathcal{W}_C) = \check{C}.$$

For an arbitrary curve C the discriminant of $\mathcal{W}_C$ will also contain the hyperplanes on $\check{\mathbb{P}}^n$ corresponding to singular points of C, and the projective subspaces of codimension two dual to the lines contained in C.

Because for a plane curve of degree k, the normal bundle of the web $\mathcal{W}_C$ is $\mathcal{O}_{\mathbb{P}^2}(2k)$, one has

$$\deg\left(\Delta(\mathcal{W}_C)\right) = \deg\left(K_{\mathbb{P}^2}^{k(k-1)} \otimes \mathcal{O}_{\mathbb{P}^2}\left(4k(k-1)\right)\right)$$
$$= -3k(k-1) + 4k(k-1) = k(k-1),$$

according to Proposition 1.3.3. In particular, one recovers the classical Plücker formula for the degree of the dual of smooth curves:

$$C \text{ smooth and } \deg(C) = k \implies \deg(\check{C}) = k(k-1).$$

If C is a plane curve with singularities, then the lines dual to the singular points will also be part of $\Delta(\mathcal{W}_C)$. The multiplicity with which these lines appear in the discriminant divisor will vary according to the analytical type of the singularity. For instance, the lines dual to ordinary nodes will appear with multiplicity two, while the lines dual to ordinary cusps will appear with multiplicity three. In particular for a degree k curve with n ordinary nodes and c ordinary cusps as singularities, one obtains another instance of Plücker formula

$$\deg(\check{C}) = k(k-1) - 2n - 3c .$$

The Monodromy of $\mathcal{W}_C$

To describe the monodromy of $\mathcal{W}_C$ for irreducible curves C the algebraic lemma below will be used.

Lemma 1.4.3. *Let G be a normal subgroup of $\mathfrak{S}_k$. If G contains a transposition, then G is the full symmetric group.*

Proof. Left as an exercise to the reader. $\qquad\square$

Proposition 1.4.4. *If C is an irreducible and non degenerate projective curve of degree k on $\mathbb{P}^n$, $n \geq 2$, then the monodromy group of $\mathcal{W}_C$ is the full symmetric group.*

Proof. Let us first consider the case when $n = 2$. Let $C \subset \mathbb{P}^2$ be an irreducible projective curve of degree $k \geq 3$ and let L be a generic line in the dual projective space $\check{\mathbb{P}}^2$. By genericity, the germ $\mathcal{W}_C(L)$ of $\mathcal{W}_C$ along L is quasi-smooth. This allows to consider the monodromy $\rho_{\mathcal{W}_C(L)} : \pi_1(L \setminus \Delta(\mathcal{W}_C), p) \to \mathfrak{S}_k$ of $\mathcal{W}_C(L)$ (for p generic in L) whose image G_L is clearly a subgroup of the full monodromy group of $\mathcal{W}_C$.

Let x be the point in $\mathbb{P}^2$ associated with L by projective duality. By genericity, x does not lie on C, thus the linear projection from x induces a surjective morphism $\pi : C \to L$. It is not difficult to see that G_L is nothing but the monodromy group of π. Let ℓ be a line tangent to C at a point c and passing through x. Still by genericity, ℓ is simply tangent to C at c and cuts $C \setminus \{c\}$ transversely. This implies that G_L contains a transposition.

Proving that G_L is normal requires more technical arguments and will not be treated here. We refer to the recent paper [116] for a proof that this is indeed the case, at least when L is generic (see Lemma 3.2 and Proposition 3.3 in [116] more specifically). According to Lemma 1.4.3, this yields $G_L = \mathfrak{S}_k$ for L generic, which implies in turn that the monodromy group of $\mathcal{W}_C$ is the full symmetric group.

Finally, the general case when $n > 2$ is dealt with a dimensional reduction to the planar case, see the proof of Theorem 3.5 in [116] for details. $\qquad\square$

Smoothness of $\mathcal{W}_C$

Proposition 1.4.5. *Let C be an irreducible non-degenerate projective curve in $\mathbb{P}^n$, $n \geq 2$. If $H \in \check{\mathbb{P}}^n$ is a generic hyperplane, then $\mathcal{W}_C(H)$ is a germ of smooth web.*

By duality, the proposition is clearly equivalent to the so-called *uniform position principle* for curves. The proof presented below follows closely [7, pp. 109–113].

Proposition 1.4.6. *If $C \subset \mathbb{P}^n$, $n \geq 2$, is an irreducible non-degenerate projective curve of degree $d \geq n$, then a generic hyperplane H intersects C at d distinct points. Moreover, any n among these d points span H.*

The restriction on the degree of C is not really a hypothesis. Indeed, every non-degenerate curve on $\mathbb{P}^n$ has degree at least n as will be shown in Proposition 2.4.11 of Chap. 2.

Proof of Propositions 1.4.5 and 1.4.6. Let $U = \check{\mathbb{P}}^n - \Delta(\mathcal{W}_C)$ and denote by $I \subset C^n \times U$ the locally closed variety defined by the relation

$$(p_1, \ldots, p_n, H) \in I \iff p_1, \ldots, p_n \text{ are distinct points in } H \cap C.$$

Because the monodromy group of $\mathcal{W}_C$ is the full symmetric group (by Proposition 1.4.4), the variety I is irreducible and in particular connected. Moreover the natural projection from I to U is surjective and has finite fibers. Therefore I has dimension $n = \dim U$.

Let now $J \subset I$ be the closed subset defined by

$$(p_1, \ldots, p_n, H) \in J \iff p_1, \ldots, p_n \text{ are contained in a } \mathbb{P}^{n-2}.$$

Since C is non-degenerated, one can choose n distinct points on it which span a $\mathbb{P}^{n-1}$. Thus J is a proper subset of I. The irreducibility of I implies $\dim J < \dim I = n$. Therefore the image of the projection to U is a proper subset, with complement parametrizing hyperplanes intersecting C with the requisite property. $\square$

1.4.3 Projective Duality

Given a global k-web $\mathcal{W}$ on $\mathbb{P}^n$ and its lift W to $\mathbb{P}T^*\mathbb{P}^n \simeq \mathcal{I}$, it is natural to inquire which sort of object W induces on $\check{\mathbb{P}}^n$ through the projection $\check{\pi}$.

To answer such a question, let us assume for a moment that $W \subset \mathcal{I}$ is irreducible or, equivalently, that the monodromy of $\mathcal{W}$ is transitive.

If the map $\check{\pi}|_W : W \to \check{\mathbb{P}}^n$ is surjective, then there exists a web $\check{\mathcal{W}}$ on $\check{\mathbb{P}}^n$ with lift to $\check{\mathcal{I}} = \mathcal{I}$ equal to W. The order of $\check{\mathcal{W}}$ is precisely the degree of $\check{\pi}|_W$, that is

$$d_0(\check{\mathcal{W}}) = d_{n-1}(\mathcal{W}).$$

In the two-dimensional case, the degree of $\check{\pi}|_W$ is nothing else than the degree of W. But beware that this is no longer true when the dimension is at least three. To determine the degree of $\check{W}$, notice that $\check{W}$ is tangent to a line ℓ at a point p if and only if one of the tangent spaces of $\check{W}$ at p contains the line ℓ. Therefore the number of tangencies of $\check{W}$ and ℓ is the intersection of W with the conormal variety $\mathrm{Con}(\ell) \subset \mathbb{P}T^*\check{\mathbb{P}}^n$ of ℓ. In other words

$$d_1(\check{W}) = d_{n-2}(W).$$

Arguing similarly, it follows that the following equalities hold true:

$$d_i(\check{W}) = d_{n-i-1}(W), \qquad i = 0, \ldots, n-1.$$

In order to deal with the case when $\check{\pi}|_W : W \to \check{\mathbb{P}}^n$ is not surjective, it is convenient to extend the definition of characteristic numbers to pairs (X, W) where $X \subset \mathbb{P}^n$ is an irreducible projective variety and W is an irreducible web[6] of codimension 1 on X. To repeat the same definition as before, all that is needed is a definition of the lift of (X, W) to $\mathbb{P}T^*\mathbb{P}^n$. Mimicking the definition of conormal variety for subvarieties of $\mathbb{P}^n$, one defines $\mathrm{Con}(X, W)$, the **conormal variety of the pair** (X, W) as the closed subvariety of $\mathbb{P}T^*\mathbb{P}^n$ characterized by the following conditions:

(a) $\mathrm{Con}(X, W)$ is irreducible;
(b) $\pi(\mathrm{Con}(X, W)) = X$;
(c) For a generic point $x \in X$, the fiber $\pi^{-1}(x) \cap \mathrm{Con}(X, W)$ is a union of linear subspaces corresponding to the projectivizations of the conormal bundles in $\mathbb{P}^n$ of the leaves of W through x.

For every pair (X, W), there exists a unique pair $\mathcal{D}(X, W)$ on $\check{\mathbb{P}}^n$ with conormal variety in $\mathbb{P}T^*\check{\mathbb{P}}^n$ equal to the conormal variety of (X, W), if the following conventions are adopted:

– an irreducible codimension one web W on $\mathbb{P}^n$ is identified with the pair $(\mathbb{P}^n, W)$;
– on an irreducible projective curve C there is only one irreducible web, the 1-web $\mathcal{P}$ whose leaves are the points of C;
– a projective curve C is identified with the pair $(C, \mathcal{P})$.

In this terminology, Proposition 1.4.2 reads as

$$\mathcal{D}(C) = W_C \quad \text{and} \quad \mathcal{D}(W_C) = C.$$

---

[6]When X is a singular variety, a web on X is a web on its smooth locus which extends to a global web on any of its desingularizations.

Notice that the pairs $(X, \mathcal{W})$ come with naturally attached characteristic numbers

$$d_i(X, \mathcal{W}) = \mathrm{Con}(X, \mathcal{W}) \cdot \mathrm{Con}(\mathbb{P}^i), \qquad i = 0, \ldots, n-1$$

and these generalize the characteristic numbers of a web $\mathcal{W}$ on $\mathbb{P}^n$ previously defined.

Example 1.4.7. Let X be an irreducible subvariety of codimension $q \geq 1$ on $\mathbb{P}^n$ and $\mathcal{W}$ an irreducible k-web on X. Since X has codimension q, for $0 \leq i \leq q-1$ a generic $\mathbb{P}^i$ linearly embedded in $\mathbb{P}^n$ does not intersect X. Therefore $d_i(X, \mathcal{W}) = 0$ for $i = 0, \ldots, q-1$. A generic $\mathbb{P}^q$ will intersect X in $\deg(X)$ smooth points and over each one of these points $\mathrm{Con}(\mathbb{P}^q)$ will intersect $\mathrm{Con}(X, \mathcal{W})$ in k points. Therefore $d_q(X, \mathcal{W}) = \deg(X) \cdot k$.

With these definitions at hand we have the following Biduality Theorem.

Theorem 1.4.8. *For any pair $(X, \mathcal{W})$, where $X \subset \mathbb{P}^n$ is an irreducible projective subvariety and $\mathcal{W}$ is an irreducible codimension one web on X, the following identity holds true*

$$\mathcal{D}\mathcal{D}(X, \mathcal{W}) = (X, \mathcal{W}).$$

Moreover the characteristic numbers of $(X, \mathcal{W})$ and $\mathcal{D}(X, \mathcal{W})$ satisfy

$$d_i(X, \mathcal{W}) = d_{n-1-i}\big(\mathcal{D}(X, \mathcal{W})\big), \qquad i = 0, \ldots, n-1.$$

Finally to deal with arbitrary pairs $(X, \mathcal{W})$, where X is not necessarily irreducible nor $\mathcal{W}$ has necessarily transitive monodromy, one writes $(X, \mathcal{W})$ as the superposition of irreducible pairs and applies $\mathcal{D}$ to each factor.

1.4.4 Webs Attached to Projective Surfaces

One particularly rich source of examples of webs on surfaces is the classical projective differential geometry widely practiced until the late thirties of the twentieth century. The simplest example is perhaps that of the asymptotic webs on surfaces in $\mathbb{P}^3$ which will now be described.

Asymptotic Webs on Surfaces in $\mathbb{P}^3$

For a classical reference on the construction and results presented below, the reader can consult [52, Chapter IV].

Let S be a germ of smooth surface in $\mathbb{P}^3$. As such it admits a parametrization $[\varphi] : (\mathbb{C}^2, 0) \to \mathbb{P}^3$ which is the projectivization of a holomorphic embedding $\varphi : (\mathbb{C}^2, 0) \to \mathbb{C}^4 \setminus \{0\}$.

For $p \in (\mathbb{C}^2, 0)$, the tangent plane of S at the point $[\varphi(p)]$ is determined by the vector subspace of $T_{\varphi(p)}\mathbb{C}^4$ generated by

$$\varphi(p), \quad \frac{\partial \varphi}{\partial x}(p) \quad \text{and} \quad \frac{\partial \varphi}{\partial y}(p).$$

A germ of smooth curve C on S admits a parametrization of the form $\varphi \circ \gamma(t)$ where $\gamma : (\mathbb{C}, 0) \to (\mathbb{C}^2, 0)$ is an immersion. Its osculating plane at $\varphi \circ \gamma(t)$ is determined by the vector space generated by $(\varphi \circ \gamma)(t)$, $(\varphi \circ \gamma)'(t)$, and $(\varphi \circ \gamma)''(t)$. Although the vectors $(\varphi \circ \gamma)'(t)$ and $(\varphi \circ \gamma)''(t)$ do depend on the choice of the parametrization $\varphi \circ \gamma$ of C, the vector space generated by them and by $(\varphi \circ \gamma)(t)$ does not.

A curve C is an **asymptotic curve** of S if, at every point p of C, its osculating plane is contained in the tangent space of S. For instance, any (piece of) line contained in S is an asymptotic curve of S.

Since $(\varphi \circ \gamma)'(t)$ always belongs to the tangent space of S at $(\varphi \circ \gamma)(t)$, being an asymptotic curve is equivalent to the vanishing of the following determinant (where each entry represents a distinct row)

$$\det\left((\varphi \circ \gamma)''(t), \varphi(\gamma(t)), \frac{\partial \varphi}{\partial x}(\gamma(t)), \frac{\partial \varphi}{\partial y}(\gamma(t))\right). \tag{1.4}$$

But for every $t \in (\mathbb{C}, 0)$, one has

$$(\varphi \circ \gamma)''(t) = D^2\varphi(\gamma(t)) \cdot \gamma'(t) \cdot \gamma'(t) + D\varphi(\gamma(t)) \cdot \gamma''(t)$$

and the image of $D\varphi(\gamma(t))$ is always contained in the vector space generated by the last three rows of the above matrix. Hence the vanishing of (1.4) is equivalent to the vanishing of

$$\det\left(D^2\varphi(\gamma(t)) \cdot \gamma'(t) \cdot \gamma'(t), \varphi(\gamma(t)), \frac{\partial \varphi}{\partial x}(\gamma(t)), \frac{\partial \varphi}{\partial y}(\gamma(t))\right).$$

This last expression can be rewritten as

$$\gamma^*\left(A\,dx^2 + 2B\,dx{\cdot}dy + C\,dy^2\right)$$

where

$$A = \det\left(\varphi_{xx}, \varphi, \varphi_x, \varphi_y\right)$$
$$B = \det\left(\varphi_{xy}, \varphi, \varphi_x, \varphi_y\right)$$
$$\text{and} \quad C = \det\left(\varphi_{yy}, \varphi, \varphi_x, \varphi_y\right).$$

It may happen that A, B, C are all identically zero. It is well-known that this is the case if and only if S is contained in a hyperplane of $\mathbb{P}^3$. It may also happen that although non-zero, the 2-symmetric differential form $A dx^2 + 2B dx dy + C dy^2$ is proportional to the square of a differential 1-form. This is the case if and only if the surface S is developable. Recall that a surface is **developable** if it is contained in a plane, a cone or the tangent surface of a curve.

In general for a non-developable surface, one gets a 2-symmetric differential form that induces an (eventually singular) 2-web on S: the **asymptotic web** of S, noted by $\mathcal{W}_S$. We let the reader verify that the latter is projectively attached to S, i.e. that $\mathcal{W}_{g(S)} = g(\mathcal{W}_S)$ for any projective automorphism $g \in PGL_4(\mathbb{C})$.

The simplest example of an asymptotic web is certainly the asymptotic web of a smooth quadric Q in $\mathbb{P}^3$. As is well-known, the latter is isomorphic to $\mathbb{P}^1 \times \mathbb{P}^1$. The fibers of both natural projections to $\mathbb{P}^1$ are lines in $\mathbb{P}^3$. Since any line contained in it is an asymptotic curve of Q, it is clear that its asymptotic web $\mathcal{W}_Q$ is formed by these two families of lines.

Asymptotic Webs: Alternative Approach

When $S \subset \mathbb{P}^3$ is a smooth projective surface the definition of the asymptotic web of S is amenable to a more intrinsic formulation. Suppose that S is cut out by an irreducible homogenous polynomial $F \in \mathbb{C}[x_0, \ldots, x_3]$. The Hessian matrix of F

$$\mathrm{Hess}(F) = \left(\frac{\partial^2 F}{\partial x_i \partial x_j} \right),$$

when evaluated at the tangent vectors of S, gives rise to a morphism

$$\mathrm{Sym}^2 TS \longrightarrow NS$$

where $NS \simeq \mathcal{O}_S(\deg(F))$ is the normal bundle of S. This morphism is usually called the (projective) **second fundamental form** of S. Dualizing it, and tensoring the result by NS one obtains a holomorphic section of $\mathrm{Sym}^2 \Omega_S^1 \otimes NS$.

When S is not developable (which under the smoothness and projectiveness assumption on S is equivalent to S not being a plane) this section, after factoring eventual codimension one components of its zero set, defines a singular 2-web on S that is nothing but the asymptotic web of S previously defined. Its discriminant coincides, set theoretically, with the locus on S defined by the vanishing of $\mathrm{Hess}(F)$.

The General Philosophy

One can abstract from the definition of asymptotic web the following procedure:

1. Take a linear system[7] $|V|$ on a surface S;
2. Consider the elements of $|V|$ with abnormal singularities at a generic point p of S;
3. If there are only finitely many abnormal elements of V for a given generic point p, consider the web with tangents at p determined by the tangent cone of these elements.

This kind of construction abounds in classical projective differential geometry.

Darboux 3-Web

Let $S \subset \mathbb{P}^3$ be a surface and let us consider the restriction to S of the linear system $|\mathcal{O}_{\mathbb{P}^3}(2)|$ of quadric surfaces in $\mathbb{P}^3$.

If S is generic enough, then at a generic point $p \in S$ there are exactly three quadrics whose restriction at S is a curve with first non-zero jet at p of the form

$$\ell_i(x, y)^3, \qquad i = 1, 2, 3$$

where (x, y) are local coordinates of S centered at p and the ℓ_i are linear forms. These three quadrics are the **quadrics of Darboux** of S at p. For more details see [81, pp. 141–144].

In this way one defines a 3-web with tangents at p given by $\mathrm{Ker}(d(\ell_i)_p) \in \mathbb{P}T_p S$ for $i = 1, 2, 3$. This is the **Darboux 3-web** of S.

Segre's 5-Web

Let now S be a surface in $\mathbb{P}^5$ and let us consider the restriction to S of the linear system of hyperplanes $|\mathcal{O}_{\mathbb{P}^5}(1)|$.

For a generic point p in a generic surface S there are exactly five hyperplanes which intersect S along a curve which has a tacnode[8] singularity at p. The five directions determined by these tacnodes are **Segre's principal directions** of S

---

[7]Recall that a linear system is the projectivization $|V| = \mathbb{P}(V)$ of a finite dimensional vector subspace $V \subset H^0(S, \mathcal{L})$, where $\mathcal{L}$ is a line-bundle on S. When S is a germ of surface, any line-bundle on it is trivial hence a linear system is nothing else than the projectivization of a finite dimensional vector space of germs of functions.

[8]An *ordinary* tacnode is a singularity of curve with exactly two branches, both of them smooth, having an ordinary tangency. Here tacnode refer to a curve cut out by a power series of the form

$$\ell(x, y)^2 + \ell(x, y)P_2(x, y) + h.o.t.$$

at p. By definition **Segre's 5-web** is defined as the 5-web determined pointwise by Segre's principal directions in the case when they are distinct at a generic point of S. In this case, one can verify that this web is projectively attached to S.

There are surfaces such that through every point there are infinitely many principal directions. For instance, the developable surfaces—planes, cones and tangent of curves—do have this property and so do the degenerated surfaces, that is surfaces contained in a proper hyperplane of $\mathbb{P}^5$. A remarkable theorem of Corrado Segre [124] says that beside these examples, the only surfaces in $\mathbb{P}^5$ with infinitely many principal directions at a generic point are the ones contained in the Veronese surface obtained through the embedding of $\mathbb{P}^2$ into $\mathbb{P}^5$ given by the linear system $|\mathcal{O}_{\mathbb{P}^2}(2)|$.

If $\varphi : (\mathbb{C}^2, 0) \to \mathbb{C}^6$ is a lift of a regular parametrization of the surface S, then the pull-back of Segre's 5-web of S is induced by the 5-symmetric differential form

$$
\omega_\varphi = \det \begin{pmatrix}
\varphi \\
\varphi_x \\
\varphi_y \\
\varphi_{xx} dx + \varphi_{xy} dy \\
\varphi_{xy} dx + \varphi_{yy} dy \\
\varphi_{xxx} dx^3 + 3\,\varphi_{xxy} dx^2 dy + 3\,\varphi_{xyy} dx dy^2 + \varphi_{yyy} dy^3
\end{pmatrix}.
$$

It can be verified that once the parametrization φ is changed by one of the form

$$
\hat{\varphi}(x, y) = \lambda(x, y) \cdot \varphi\big(\psi(x, y)\big)
$$

where λ is a unit in $\mathcal{O}_{(\mathbb{C}^2, 0)}$ and $\psi \in \mathrm{Diff}(\mathbb{C}^2, 0)$ is a germ of biholomorphism, then one has

$$
\omega_{\hat{\varphi}} = \lambda^6 \cdot \det(D\psi)^2 \cdot \psi^*(\omega_\varphi).
$$

This identity implies that the collection $\{\omega_\varphi\}$, with φ ranging over germs of parametrizations of S, defines a holomorphic section of

$$
\mathrm{Sym}^5\big(\Omega_S^1\big) \otimes \mathcal{O}_S(6) \otimes K_S^{\otimes 2}.
$$

A nice example is given by the surface S obtained as the image of the rational map from $\mathbb{P}^2$ to $\mathbb{P}^5$ determined by the linear system of cubics passing through four points $p_1, p_2, p_3, p_4 \in \mathbb{P}^2$ in general position.[9] At a generic point $p \in \mathbb{P}^2$, the five

---

where ℓ is a linear form and P_2 is a homogeneous form of degree 2.

[9]This surface is a Del Pezzo surface of degree 5.

cubics in the linear system with a tacnode at p are: the union of the conic through p_1, p_2, p_3, p_4 and p with its tangent line at p; and for every $i \in \underline{4}$, the union of the line $\overline{pp_i}$ with the conic through p and all the p_j with $j \neq i$ which is moreover tangent to $\overline{pp_i}$ at p. This geometric description clearly shows that Segre's 5-web of S is nothing else than Bol's 5-web $\mathcal{B}_5$ presented in Sect. 1.2.3.

Chapter 2
Abelian Relations

In this chapter, the key concept of abelian relation is introduced. Roughly speaking, an abelian relation for a web $\mathcal{W}$ is an additive functional equation among the first integrals of its foliations. The credo here is that the notion of abelian relation must be considered as a generalization for webs of the more classical notion of holomorphic differential form on a projective variety. The most compelling evidences supporting this credo will be presented in the next two chapters through the study of abelian relations of algebraic webs.

The abelian relations of a germ of quasi-smooth web $\mathcal{W}$ form a complex vector space denoted by $\mathcal{A}(\mathcal{W})$. One of the main goals of this chapter is to prove that $\mathcal{A}(\mathcal{W})$ is indeed a finite dimensional vector space, and that when $\mathcal{W}$ is smooth its dimension—the rank of $\mathcal{W}$—is bounded by a finite explicit number that depends only on n and on k.

This bound was established by Bol when $n = 2$, and by Chern in his Ph.D. thesis under the direction of Blaschke when n is arbitrary. Before embarking upon its proof, carried out in Sect. 2.3, we will discuss the determination of the space of abelian relations for planar webs in some particular cases in Sect. 2.2.

Of tantalizing importance for what is to come later in Chap. 5 is the content of Sect. 2.4. There Castelnuovo's results on the geometry of point sets in projective spaces are proved, and from them are deduced constraints on the geometry of webs of maximal rank, that is webs whose rank attains Chern's bound.

2.1 Definition

Let $\mathcal{W}$ be a quasi-smooth k-web on $(\mathbb{C}^n, 0)$. As said above, an abelian relation for $\mathcal{W}$ is an additive functional equation among the first integrals of the foliations defining $\mathcal{W}$.

© Springer International Publishing Switzerland 2015

J.V. Pereira, L. Pirio, *An Invitation to Web Geometry*, IMPA Monographs 2,
DOI 10.1007/978-3-319-14562-4_2

More precisely, if $W = W(\omega_1, \ldots, \omega_k)$ for some germs of holomorphic 1-forms $\omega_1, \ldots, \omega_k$ on $(\mathbb{C}^n, 0)$, then an **abelian relation** of W is a k-uple $(\eta_i)_{i=1}^k$ of germs of 1-forms $\eta_i \in \Omega^1(\mathbb{C}^n, 0)$, $i \in \underline{k}$, satisfying the following three conditions:

(a) for every $i \in \underline{k}$, the 1-form η_i is closed, that is $d\eta_i = 0$;
(b) for every $i \in \underline{k}$, the 1-form η_i is trivial along the leaves of $\mathcal{F}_i$, that is $\omega_i \wedge \eta_i = 0$;
(c) the 1-forms η_i sum up to zero, that is $\sum_{i=1}^k \eta_i = 0$.

Notice that a primitive of η_i exists, since η_i is closed. Such a primitive is a first integral of $\mathcal{F}_i$, because $\omega_i \wedge \eta_i = 0$. In particular, if $\mathcal{F}_i$ is defined through a submersion $u_i : (\mathbb{C}^n, 0) \to (\mathbb{C}, 0)$, then

$$\int_0^z \eta_i = g_i(u_i(z)),$$

for some germs of holomorphic functions $g_i : (\mathbb{C}, 0) \to (\mathbb{C}, 0)$. Condition (c) translates into

$$\sum_{i=1}^k g_i \circ u_i = 0$$

which is the functional equation among the first integrals of W mentioned at the beginning of the discussion.

It will be convenient to designate by $\sum_{i=1}^k \eta_i = 0$ an abelian relation $(\eta_i)_{i=1}^k \in \Omega^1(\mathbb{C}^n, 0)^k$. It is said to be **non-trivial** if at least one of the η_i's is not identically zero. If none of the 1-forms η_i is identically zero, then the abelian relation $\sum_{i=1}^k \eta_i = 0$ is called **complete**.

With the concept of abelian relation at hand, Theorem 1.2.4 can be rephrased as the following equivalences for a germ W of smooth 3-web on $(\mathbb{C}^2, 0)$:

$$W \text{ is hexagonal} \iff K(W) \equiv 0 \iff \begin{array}{l} W \text{ has a non-trivial} \\ \text{abelian relation.} \end{array}$$

To some extent, the main results presented in this book can be seen as generalizations of this equivalence to arbitrary webs of codimension one.

It is clear from the definition of an abelian relation that for a given germ of quasi-smooth k-web W the set of all abelian relations of W forms a $\mathbb{C}$-vector space, the **space of abelian relations of** W, which will be denoted by $\mathcal{A}(W)$.

If $W = W(\omega_1, \ldots, \omega_k)$, then one can write

$$\mathcal{A}(W) = \left\{ (\eta_i)_{i=1}^k \in \left(\Omega^1(\mathbb{C}^n, 0)\right)^k \;\middle|\; \text{for every } i \in \underline{k} : \begin{array}{l} d\eta_i = 0 \\ \omega_i \wedge \eta_i = 0 \\ \sum_{i=1}^k \eta_i = 0 \end{array} \right\}.$$

Notice that a germ of diffeomorphism φ establishing an equivalence between two germs of k-webs $\mathcal{W}$ and $\mathcal{W}' = \varphi_*(\mathcal{W})$ induces a natural linear isomorphism between their spaces of abelian relations:

$$\mathcal{A}(\mathcal{W}') \longrightarrow \mathcal{A}(\mathcal{W})$$

$$(\eta_i')_{i=1}^k \longmapsto \left(\varphi^*(\eta_i')\right)_{i=1}^k.$$

Consequently, the dimension of $\mathcal{A}(\mathcal{W})$ is an analytic invariant attached to $\mathcal{W}$.

We will prove below in Sect. 2.3 that when $\mathcal{W}$ is a germ of smooth k-web on $(\mathbb{C}^n, 0)$ the space of abelian relations $\mathcal{A}(\mathcal{W})$ is indeed a finite dimensional vector space and that its dimension, the **rank of** $\mathcal{W}$, denoted by

$$\mathrm{rank}(\mathcal{W}) = \dim\left(\mathcal{A}(\mathcal{W})\right),$$

is bounded from above by **Castelnuovo's number** defined as

$$\pi(n,k) = \sum_{j=1}^{\infty} \max\left(0, k - j(n-1) - 1\right).$$

The space of abelian relations of a global web is no longer a vector space but a local system defined on an open subset containing the complement of the discriminant of the web. This can be inferred from the results by Pantazi–Hénaut expounded in Sect. 6.3 of Chap. 6. For an elementary and simple argument, see [107, Théorème 1.2.2]. Note that both the approaches mentioned above deal a priori with webs on surfaces, but deducing the general case from them can be done without any real difficulty.

2.2 Determining the Abelian Relations

As mentioned above, if $\mathcal{W}$ is a quasi-smooth k-web on $(\mathbb{C}^n, 0)$ and if $\varphi : (\mathbb{C}^2, 0) \rightarrow (\mathbb{C}^n, 0)$ is a generic holomorphic immersion, then $\varphi^*\mathcal{W}$ is a smooth k-web on $(\mathbb{C}^2, 0)$ and φ induces an injection of $\mathcal{A}(\mathcal{W})$ into $\mathcal{A}(\varphi^*\mathcal{W})$. Thus, the specialization to the two-dimensional case, which will be in use up to the end of this section, is not seriously restrictive.

2.2.1 Abel's Method

Before the first appearance of web geometry, Abel already studied functional equations of the form

$$\sum_{i=1}^{k} g_i \circ u_i = 0$$

for given functions u_i depending on two variables. In his first published paper [1], he devised a method to determine the functions g_i satisfying this functional equation. Abel's method will not be presented here in its full generality but the particular case when all but one of the functions u_i are homogenous polynomials of degree one will be scrutinized following [113]. Under these additional assumptions, Abel's method is remarkably simplified but still leads to interesting examples of functional equations and webs. For a comprehensive account and modern exposition of Abel's method in the context of web geometry, the reader can consult [109].

For $i \in \underline{k}$, let $u_i(x, y) = a_i x + b_i y$ where $a_i, b_i \in \mathbb{C}$ are complex numbers satisfying $a_i b_j - a_j b_i \neq 0$ whenever $i \neq j$. These conditions imply the smoothness of the k-web $\mathcal{W}(u_1, \ldots, u_k)$. It will be convenient to consider the vector fields $v_i = b_i \partial_x - a_i \partial_y$ which define the very same foliation as the submersions u_i.

Let $u : (\mathbb{C}^2, 0) \to (\mathbb{C}, 0)$ be a germ of holomorphic submersion satisfying $v_i(u) \neq 0$ for every $i \in \underline{k}$, and let us consider the smooth $(k + 1)$-web on $(\mathbb{C}^2, 0)$:

$$\mathcal{W} = \mathcal{W}(u_1, \ldots, u_k, u).$$

To determine the rank of $\mathcal{W}$, it suffices to look for holomorphic solutions $g, g_1, \ldots, g_k : (\mathbb{C}, 0) \to (\mathbb{C}, 0)$ of the functional equation

$$g \circ u = \sum_{i=1}^{k} g_i \circ u_i .$$

For this purpose, let us apply the derivation v_1 to both sides of the equation above to obtain

$$g'(u) \cdot v_1(u) = \sum_{i=2}^{k} g_i{}'(u_i) \cdot v_1(u_i) .$$

Notice that u_1 no longer appears in the right-hand side.

Let us now apply the derivation v_2 to this new equation. We use the commutativity of v_1 and v_2, that is $[v_1, v_2] = 0$, to get

$$g''(u) \cdot v_2(u) \cdot v_1(u) + g'(u) \cdot v_2\big(v_1(u)\big)$$

$$= \sum_{i=3}^{k} g_i{}''(u_i) \cdot v_2(u_i) \cdot v_1(u_i) + (g_i)'(u_i) \cdot v_2\big(v_1(u_i)\big) .$$

Iterating this procedure, one obtains an equation of the form

$$\left(\prod_{i=1}^{k} v_i(u)\right) g^{(k)}(u) + \cdots + v_k\big(v_{k-1}(\cdots v_1(u))\big) g'(u) = 0$$

which after dividing by the coefficient of $g^{(k)}(u)$ (that does not vanish since $\mathcal{W}$ is smooth by hypothesis) can be written

$$g^{(k)}(u) = \sum_{i=1}^{k-1} h_i \, g^{(i)}(u)$$

where the h_i's are germs of meromorphic functions.

Let $v = u_y \partial_x - u_x \partial_y$ be the hamiltonian vector field of u. If for some $i \in \underline{k}$ the function $v(h_i)$ is not identically zero, then one can apply the derivation v to the above equation in order to reduce its order. Otherwise the functions h_i are functions of u only, that is $h_i = h_i(u)$ for every $i \in \underline{k}$.

Eventually one obtains a linear differential equation of the form

$$g^{(\ell)}(u) = \sum_{i=1}^{\ell-1} h_i(u) g^{(i)}(u) \tag{2.1}$$

with $\ell \le k$. Thus the possibilities for g are reduced to a finite dimensional vector space: the space of solutions of (2.1).

After discarding the constant solutions of (2.1) one notices at this point that

$$\mathrm{rank}\,(\mathcal{W}) = \mathrm{rank}\,(\mathcal{W}(u_1,\ldots,u_k)) + k - 1$$

when $u_1,\ldots,u_k$ are linear homogeneous polynomials.

Beware that this is no longer true if the linear polynomials u_i are replaced by arbitrary submersions, the point being that the hamiltonian vector fields v_i no longer commute. One can still work his way out to deduce that an equation such as (2.1) will still hold true, as it is done in [109], but it will be no longer true that ℓ is bounded by k.

Example 2.2.1. Let $u_1,\ldots,u_k \in \mathbb{C}[x,y]$ be homogeneous linear polynomials. Suppose that they are pairwise linearly independent and let $\mathcal{W} = \mathcal{W}(u_1,\ldots,u_k)$ be the induced k-web. Then

$$\mathrm{rank}(\mathcal{W}) = \frac{(k-1)(k-2)}{2}$$

Proof. The proof goes by induction. For $k = 2$ there is no abelian relation. Suppose that the result holds true for $k \ge 2$, that is every parallel k-web has rank $(k-1)(k-2)/2$. Looking for solutions of

$$g \circ u_{k+1} = \sum_{i=1}^{k} g_i \circ u_i$$

following the strategy described above one obtains the equation $g^{(k)}(u_{k+1}) = 0$. Thus g must be a polynomial in $\mathbb{C}[t]$ of degree at most $(k-1)$. Imposing that $g(0) = 0$ leaves a vector space of dimension $k - 2$ to choose g from. Hence the rank of $\mathcal{W}_{k+1} = \mathcal{W}(u_1, \ldots, u_{k+1})$ is bounded by $(k-1)(k-2)/2 + (k-2)$.

But for every positive integer $j \leq k-1$, a dimension count shows that $(u_{k+1})^j = \sum_{i=1}^{k} \lambda_{i,j} \cdot (u_i)^j$ for suitable $\lambda_{i,j} \in \mathbb{C}$. It follows that the rank of $\mathcal{W}_{k+1}$ is $k(k-1)/2$ as requested.   □

In the particular case under analysis one can make use of the following lemma.

Lemma 2.2.2. *Suppose as above that the functions u_i are linear homogenous and, again as above, let v_i be the hamiltonian vector field of u_i. The following assertions are equivalent:*

(a) the function $g(u)$ is of the form $\sum_{i=1}^{k} g_i(u_i)$;
(b) the identity $v_1 \cdot v_2 \cdots v_k\big(g(u)\big) \equiv 0$ holds true.

Proof. Clearly (a) implies (b). The converse will be proved by induction. For $k = 1$ the result is evident. By induction hypothesis,

$$v_k\big(g(u)\big) = \sum_{i=1}^{k-1} h_i(u_i).$$

If H_i is a primitive of h_i, then $v_k(H_i(u_i)) = h_i(u_i) \cdot v_k(u_i)$. Because $v_k(u_i)$ is a non-zero constant when $i < k$, one can write

$$v_k\left(g(u) - \sum_{i=1}^{k-1} v_k(u_i)^{-1} H_i(u_i)\right) = 0.$$

To conclude it suffices to apply the basis of the induction.   □

Proposition 2.2.3. *Let $\mathcal{W}$ be as in Example 2.2.1. Let also $u : (\mathbb{C}^2, 0) \to (\mathbb{C}, 0)$ be a submersion and let $\mathcal{F}$ be the induced foliation. If the $(k+1)$-web $\mathcal{W} \boxtimes \mathcal{F}$ is smooth, then*

$$\mathrm{rank}(\mathcal{W} \boxtimes \mathcal{F}) \leq \frac{k(k-1)}{2}.$$

Moreover, the equality holds true if and only if $\ell = k$ in Eq. (2.1).

Proof. First notice that Eq. (2.1) has been derived by first developing formally

$$v_1 \cdot v_2 \cdots v_k \big(g(u)\big) = \sum_{i=1}^{k} f_i(x, y) g^{(i)}(u),$$

then dividing by f_k, setting $h_i = f_i/f_k$ and deriving repeatedly with respect to the hamiltonian vector field of u until arriving to an equation depending only on u. Therefore, Lemma 2.2.2 implies that any solution g of

$$g(u) = \sum_{i=1}^{k} g_i(u_i)$$

will also be a solution of (2.1). But if $\mathrm{rank}(\mathcal{W} \boxtimes \mathcal{F}) = \frac{k(k-1)}{2}$ then Eq. (2.1) has to have at least $k - 1$ non-constant solutions vanishing at zero. Consequently $k = \ell$.

Reciprocally if $k = \ell$ then there will be $k - 1$ non-constant solutions vanishing at zero for $v_1 v_2 \cdots v_k(g(u)) = 0$. Lemma 2.2.2 implies that $\mathrm{rank}(\mathcal{W} \boxtimes \mathcal{F}) = \mathrm{rank}(\mathcal{W}) + (k - 1)$.   □

Webs such as $\mathcal{W} \boxtimes \mathcal{F}$ of the proposition above, which are obtained from the superposition of a parallel web and one non-linear foliation, will be called **quasi-parallel** webs.

Example 2.2.4. Let $u : (\mathbb{C}^2, 0) \to \mathbb{C}$ be a submersion of the form $u(x, y) = a(x) + b(y)$ with $a, b \in \mathcal{O}(\mathbb{C}, 0)$. Assume that the quasi-parallel 5-web $\mathcal{W}(x, y, x - y, x + y, u(x, y))$ is smooth, that is $a_x b_y (a_x^2 - b_y^2)(0) \neq 0$. A straightforward computation shows that

$$v_1 v_2 v_3 v_4 \big(g(u)\big) = g''''(u) a_x b_y (a_x^2 - b_y^2)$$

$$+ 3g'''(u) a_x b_y (a_{xx} - b_{yy}) + g''(u)(b_y a_{xxx} - a_x b_{yyy}).$$

Proposition 2.2.3 implies that $\mathcal{W}$ has rank equal to 6 if and only if

$$v_u \left(\frac{a_{xx} - b_{yy}}{a_x^2 - b_y^2} \right) = 0 \quad \text{and} \quad v_u \left(\frac{a_y b_{xxx} - a_x b_{yyy}}{a_x b_y (a_x^2 - b_y^2)} \right) = 0$$

where $v_u = b_y \partial_x - a_x \partial_y$.

The simplest functions $u(x, y) = a(x) + b(y)$ satisfying this system of partial differential equations are $x^2 + y^2$, $x^2 - y^2$, $\exp x + \exp y$, $\log(\sin x \sin y)$ and $\log(\tanh x \tanh y)$. But there are still other functions. There is a continuous family of solutions that can be written with the help of theta functions of elliptic curves.

The 5-webs of the form $\mathcal{W}(x, y, x - y, x + y, a(x) + b(y))$ with rank equal to 6 have been completely classified in [113] through a careful analysis of the above system of PDEs.

If nothing else, this example shows how involved the search for webs of high rank can be, even in considerably simple cases.

2.2.2 Webs with Infinitesimal Automorphisms

Let $\mathcal{F}$ be a germ of smooth foliation on $(\mathbb{C}^2, 0)$ induced by a germ of 1-form ω. A germ of vector field v is an **infinitesimal automorphism of** $\mathcal{F}$ if the foliation $\mathcal{F}$ is preserved by the local flow of v. In algebraic terms: $L_v(\omega) \wedge \omega = 0$ where $L_v = i_v \circ d + d \circ i_v$ is the **Lie derivative** with respect to the vector field v. Those who are not familiar with the Lie derivative can find its basic algebraic properties in [58, Chapter IV].

When the infinitesimal automorphism v is transverse to $\mathcal{F}$, that is when $\omega(v) \neq 0$, then an elementary computation (see [104, Corollary 2] or [29, Chapter III Section 2]) shows that the 1-form

$$\alpha = \frac{\omega}{i_v(\omega)}$$

is closed and satisfies $L_v(\alpha) = 0$. By definition, the integral

$$u(z) = \int_0^z \alpha$$

is the **canonical first integral** of $\mathcal{F}$ with respect to v. Clearly $u(0) = 0$ and $L_v(u) = v(u) = 1$. In particular the latter equality implies that u is a germ of submersion.

The canonical first integral admits a nice *physical* interpretation: its value at z measures the time taken by the local flow of v to move the leaf through zero to the leaf through z.

Now let $\mathcal{W} = \mathcal{W}(\omega_1, \ldots, \omega_k)$ be a germ of smooth k-web on $(\mathbb{C}^2, 0)$ and let v be an **infinitesimal automorphism of** $\mathcal{W}$, in the sense that v is an infinitesimal automorphism of all the foliations defining $\mathcal{W}$.

By hypothesis, one has $L_v \omega_i \wedge \omega_i = 0$ for every $i \in \underline{k}$. Because L_v commutes with d, it induces a linear map

$$L_v : \mathcal{A}(\mathcal{W}) \longrightarrow \mathcal{A}(\mathcal{W}) \tag{2.2}$$

$$(\eta_1, \ldots, \eta_k) \longmapsto (L_v \eta_1, \ldots, L_v \eta_k).$$

A simple analysis of the L_v-invariant subspaces of $\mathcal{A}(\mathcal{W})$ will provide valuable information about the abelian relations of webs admitting infinitesimal automorphisms.

Description of $\mathcal{A}(\mathcal{W})$

Suppose that $\mathcal{W} = \mathcal{F}_1 \boxtimes \cdots \boxtimes \mathcal{F}_k$ is a smooth k-web on $(\mathbb{C}^2, 0)$ which admits an infinitesimal automorphism v, regular and transverse to all the foliations $\mathcal{F}_i$.

Let $i \in \underline{k}$ be fixed. Set $\mathcal{A}_i(\mathcal{W})$ as the vector subspace of $\Omega^1(\mathbb{C}^2, 0)$ spanned by the i-th components η_i of abelian relations $(\eta_1, \ldots, \eta_k) \in \mathcal{A}(\mathcal{W})$. In other words, if $p_i : \Omega^1(\mathbb{C}^2, 0)^k \to \Omega^1(\mathbb{C}^2, 0)$ is the projection onto the i-th factor, then $\mathcal{A}_i(\mathcal{W}) = p_i(\mathcal{A}(\mathcal{W}))$.

If $u_i = \int \alpha_i$ is the canonical first integral of $\mathcal{F}_i$ with respect to v, then for $\eta_i \in \mathcal{A}_i(\mathcal{W})$, there exists a germ $f_i \in \mathcal{O}(\mathbb{C}, 0)$ for which $\eta_i = f_i(u_i)\, du_i$.

Assume now that $\mathcal{A}_i(\mathcal{W})$ is not the zero vector space and let

$$\{\eta_i^\mu = f_\mu(u_i)\, du_i \mid \mu \in \underline{n_i}\}$$

be a basis of it, consequently $n_i = \dim \mathcal{A}_i(\mathcal{W})$. Since $L_v : \mathcal{A}_i(\mathcal{W}) \to \mathcal{A}_i(\mathcal{W})$ is a linear map, there exist complex constants c_{iv}^μ, for $\mu, v = 1, \ldots, n_i$, such that

$$L_v(\eta_i^\mu) = \sum_{v=1}^{n_i} c_{iv}^\mu \, \eta_i^v, \qquad \mu \in \underline{n_i}. \tag{2.3}$$

But for any $\mu \in \underline{n_i}$, the identity below holds true

$$L_v(\eta_i^\mu) = L_v\big(f_\mu(u_i)\, du_i\big)$$
$$= v\big(f_\mu(u_i)\big) du_i + f_\mu(u_i)\, L_v\big(du_i\big) = f_\mu'(u_i)\, du_i.$$

Thus the relations (2.3) are equivalent to the following:

$$f_\mu' = \sum_{v=1}^{n_i} c_{iv}^\mu \, f_v, \qquad \mu \in \underline{n_i}. \tag{2.4}$$

Now let $\lambda_1, \ldots, \lambda_\tau \in \mathbb{C}$ be the eigenvalues of the map L_v acting on $\mathcal{A}(\mathcal{W})$ corresponding to Jordan blocks of respective dimensions $\sigma_1, \ldots, \sigma_\tau$. The system of linear differential equations (2.4) provides the following description of $\mathcal{A}(\mathcal{W})$.

Proposition 2.2.5. *The abelian relations of $\mathcal{W}$ are of the form*

$$P_1(u_1)\, e^{\lambda_i u_1}\, du_1 + \cdots + P_k(u_k)\, e^{\lambda_i u_k}\, du_k = 0$$

where $P_1, \ldots, P_k$ are polynomials of degree less or equal to σ_i.

Proposition 2.2.5 suggests an effective approach to determine $\mathcal{A}(\mathcal{W})$. Once the possible non-zero eigenvalues of the map (2.2) are restricted to a finite set then the abelian relations can be found by simple linear algebra.

To restrict the possible eigenvalues, first notice that 0 is an eigenvalue of (2.2) if and only if for every germ of vector field w the Wronskian determinant

$$\det \begin{pmatrix} u_1 & \cdots & u_k \\ w(u_1) & \cdots & w(u_k) \\ \vdots & \ddots & \vdots \\ w^{k-1}(u_1) & \cdots & w^{k-1}(u_k) \end{pmatrix} \tag{2.5}$$

is identically zero.[1] In fact, if this is the case, then there are two possibilities: either the functions $u_1, \ldots, u_k$ are $\mathbb{C}$-linearly dependent or all the orbits of w are cut out by some element of the linear system generated by $u_1, \ldots, u_k$, see [103, Theorem 4]. In particular, if w is a vector field of the form $w = \mu x \frac{\partial}{\partial x} + y \frac{\partial}{\partial y}$, with $\mu \in \mathbb{C} \setminus \mathbb{Q}$, then the leaves of w accumulate at 0 and not cut out by any holomorphic function. Therefore the vanishing of (2.5) implies the existence of an abelian relation of the form

$$\sum c_i u_i = 0,$$

where the c_i's are complex constants.

To determine the possible complex numbers λ which are eigenvalues of the map (2.2) first notice that any such eigenvalue corresponds to a functional equation of the form $c_1 e^{\lambda u_1} + \cdots + c_k e^{\lambda u_k} = \text{cst.}$ where, as before, the c_i's are complex constants. In the same spirit of what has just been made for the zero eigenvalue case consider the holomorphic function given by

$$\det \begin{pmatrix} \exp(\lambda u_1) & \cdots & \exp(\lambda u_k) \\ w\big(\exp(\lambda u_1)\big) & \cdots & w\big(\exp(\lambda u_k)\big) \\ \vdots & \ddots & \vdots \\ w^{k-1}\big(\exp(\lambda u_1)\big) & \cdots & w^{k-1}\big(\exp(\lambda u_k)\big) \end{pmatrix} \tag{2.6}$$

for an arbitrary germ of vector field w.

The Wronskian determinant (2.6) is of the form

$$\exp\big(\lambda(u_1 + \cdots + u_k)\big)\lambda^{k-1} P_w(\lambda),$$

where P_w is a polynomial in λ, of degree at most $\frac{(k-1)(k-2)}{2}$, with germs of holomorphic functions as coefficients. The common constant roots of these polynomials, when w varies, are exactly the eigenvalues of the map (2.2).

---

[1]Here, for $n \in \mathbb{N}$, w^n stands for the n-th differential operator obtained by applying n times w: inductively, one has $w^n(u) = w(w^{n-1}(u))$ for any holomorphic germ u, with the convention that $w^0 = \text{Id}$.

Example 2.2.6. The k-web $\mathcal{W}$ induced by the functions $f_i(x, y) = y + x^i$, $i = 1, \ldots, k$, has no abelian relations.

Proof. Notice that the vector field $v = \frac{\partial}{\partial y}$ is an infinitesimal automorphism of $\mathcal{W}$ and $v(df_i) = 1$ for every $i \in \underline{k}$. It follows that the f_i's are the canonical first integrals of $\mathcal{W}$. On the other hand, $P_w(\lambda)|_{x=y=0} = (-1)^{k-1} \prod_{n=1}^{k-1} n!$ for the vector field $w = \frac{\partial}{\partial x}$. Consequently, the only candidate for an eigenvalue of the map (2.2) is $\lambda = 0$. Because the functions f_i are linearly independent over $\mathbb{C}$, the web $\mathcal{W}$ carries no abelian relation at all. □

In the next example, one determines the abelian relations of one of the 5-webs of rank 6 discussed in Example 2.2.4.

Example 2.2.7. The radial vector field $R = x\frac{\partial}{\partial x} + y\frac{\partial}{\partial y}$ is an infinitesimal automorphism of the 5-web $\mathcal{W} = \mathcal{W}(x, y, x + y, x - y, x^2 + y^2)$. The canonical first integrals are $u_1 = \log x$, $u_2 = \log y$, $u_3 = \log(x + y)$, $u_4 = \log(x - y)$ and $u_5 = \frac{1}{2}\log(x^2 + y^2)$.

If $w = x\frac{\partial}{\partial x} - y\frac{\partial}{\partial y}$, then the polynomial $P_w(\lambda)$ is a complex multiple of

$$x^7 y^7 \lambda (\lambda - 1)^2 (\lambda - 2)^2 (\lambda - 4)(\lambda - 6).$$

According to Proposition 2.2.5, it suffices to look for abelian relations of the form

$$\sum_{i=1}^{5} P_{i\lambda}(\log f_i) f_i^\lambda \frac{df_i}{f_i} = 0,$$

for $\lambda = 0, 1, 2, 4, 6$, where $P_{i\lambda}$ are polynomials and $f_i = \exp(u_i)$.

Looking first for abelian relations where the polynomials $P_{i\lambda}$ are constant polynomials one has to find linear dependencies between the linear polynomials $f_1, \ldots, f_4$ and between the degree λ polynomials $f_1^\lambda, \ldots, f_4^\lambda$ and $f_5^{\lambda/2}$ for $\lambda = 2, 4, 6$.

For $\lambda = 1$ there are two linearly independent abelian relations:

$$f_1 + f_2 - f_3 = 0 \qquad \text{and} \qquad f_1 - f_2 - f_4 = 0.$$

For $\lambda = 2$ there are another two:

$$f_1^2 + f_2^2 - f_5 = 0 \qquad \text{and} \qquad f_3^2 + f_4^2 - 2f_5 = 0.$$

Finally, there is one abelian relation for $\lambda = 4$ and another one for $\lambda = 6$:

$$4f_1^4 + 4f_2^4 + f_3^4 + f_4^4 - 6f_5^2 = 0, \qquad 8f_1^6 + 8f_2^6 + f_3^6 + f_4^6 - 10f_5^3 = 0.$$

According to Proposition 2.2.3, rank($\mathcal{W}$) ≤ 6. Since the six abelian relations above are linearly independent, they form a basis of $\mathcal{A}(\mathcal{W})$.

2.3   Bounds for the Rank

Let $\mathcal{W} = \mathcal{F}_1 \boxtimes \cdots \boxtimes \mathcal{F}_k = \mathcal{W}(\omega_1, \cdots, \omega_k)$ be a germ of quasi-smooth k-web on $(\mathbb{C}^n, 0)$.

For every positive integer j, define $\mathcal{L}^j(\mathcal{W})$ as the vector subspace of the $\mathbb{C}$-vector space $\mathrm{Sym}^j \Omega_0^1(\mathbb{C}^n, 0)$ generated by $\{\omega_i(0)^j \, | \, i \in \underline{k}\}$, the j-th symmetric powers of the constant differential forms $\omega_i(0)$. Set

$$\ell^j(\mathcal{W}) = \dim \mathcal{L}^j(\mathcal{W}).$$

Equivalently, one can define $\ell^j(\mathcal{W})$ in terms of the linear parts of the submersions $u_i : (\mathbb{C}^n, 0) \to (\mathbb{C}, 0)$ defining $\mathcal{W}$. If h_i is the linear part at the origin of u_i, then

$$\ell^j(\mathcal{W}) = \dim \left(\mathbb{C}(h_1)^j + \cdots + \mathbb{C}(h_k)^j \right). \tag{2.7}$$

where $\mathbb{C}(h_1)^j + \cdots + \mathbb{C}(h_k)^j$ stands for the subspace of the space of degree j homogeneous polynomials on $\mathbb{C}^n$ generated by the j-th powers of the h_i's.

2.3.1   Bounds for $\ell^j(\mathcal{W})$

From (2.7), it follows that for any j, the integer $\ell^j(\mathcal{W})$ is bounded from above by the dimension of the space of degree j homogeneous polynomials in n variables, that is

$$\ell^j(\mathcal{W}) \leq \min \left(k, \binom{n+j-1}{n-1} \right). \tag{2.8}$$

A good lower bound is more delicate to obtain. For smooth webs there is the following proposition.

Proposition 2.3.1. *If $\mathcal{W}$ is a germ of smooth k-web on $(\mathbb{C}^n, 0)$, then*

$$\ell^j(\mathcal{W}) \geq \min \left(k, j(n-1) + 1 \right).$$

The key point is next lemma which translates questions about the dimension of vector spaces generated by powers of linear forms into questions about the codimension of space of hypersurfaces containing finite sets of points.

Lemma 2.3.2. *Let $h_1, \ldots, h_k \in \mathbb{C}_1[x_1, \ldots, x_n]$ be linear forms and let $\mathcal{P} = \{[h_1], \ldots, [h_k]\}$ be the corresponding set of points of $\mathbb{P}^{n-1} = \mathbb{P}\mathbb{C}_1[x_1, \ldots, x_n]$. If $V(j) \subset |\mathcal{O}_{\mathbb{P}^{n-1}}(j)|$ is the linear system of degree j hypersurfaces through $\mathcal{P}$, then*

$$\dim\left(\mathbb{C}(h_1)^j + \cdots + \mathbb{C}(h_k)^j\right) = \dim |\mathcal{O}_{\mathbb{P}^{n-1}}(j)| - \dim V(j).$$

Proof. For $j \in \underline{k}$, set $n_j = h^0(\mathbb{P}^{n-1}, \mathcal{O}_{\mathbb{P}^{n-1}}(j)) - 1$ and consider the j-th Veronese embedding

$$v_j : \mathbb{P}^{n-1} \longrightarrow \mathbb{P}^{n_j} = \mathbb{P}H^0\left(\mathbb{P}^{n-1}, \mathcal{O}_{\mathbb{P}^{n-1}}(j)\right)$$

$$[h] \longmapsto [h^j].$$

On the one hand, the projective dimension of $\mathbb{C}(h_1)^j + \cdots + \mathbb{C}(h_k)^j$ is equal to the dimension of the linear span of the image of $\mathcal{P}$. On the other hand, the codimension of this linear span is equal to the dimension of the linear system of hyperplanes containing it. But the pull-backs under v_j of these hyperplanes are exactly the elements of $|V(j)|$, the degree j hypersurfaces in $\mathbb{P}^{n-1}$ containing $\mathcal{P}$. The lemma follows.   □

Proof of Proposition 2.3.1. Let, as above, h_i be the linear parts of the submersions defining $\mathcal{W}$ and let $\mathcal{P} \subset \mathbb{P}^{n-1}$ be the set of k points in general position determined by the linear forms $h_1, \ldots, h_k$. According to the preceding lemma, proving the proposition only needs that $\mathcal{P}$ imposes $m = \min(k, j(n-1)+1)$ independent conditions on the space of degree j hypersurfaces in $\mathbb{P}^{n-1}$. For that sake, it suffices to show that for a subset $\mathcal{Q} \subset \mathcal{P}$ of cardinality m one can construct for each $q \in \mathcal{Q}$ a degree j hypersurface that passes through all the points of $\mathcal{Q}$ except q.

The set $\mathcal{Q} - \{q\}$ can be written as a disjoint union of j subsets of cardinality at most $(n-1)$. Any of these subsets can be supposed contained in a hyperplane that does not contain q. Thus there exists a union of j hyperplanes that contains $\mathcal{Q} - \{q\}$ and avoids q.   □

Remark 2.3.3. Notice that the proof of Proposition 2.3.1 shows that when $k \leq j(n-1)+1$ any k points in general position impose k independent conditions on the linear system of degree j hypersurfaces on $\mathbb{P}^{n-1}$.

For an essentially equivalent proof of Proposition 2.3.1, but with a more analytic flavor, see [130, Lemme 2.1].

Corollary 2.3.4. *If $\mathcal{W}$ is a germ of smooth k-web on $(\mathbb{C}^2, 0)$, then*

$$\ell^j(\mathcal{W}) = \min(k, j+1).$$

Proof. One has just to observe that the space of homogenous polynomials of degree j in two variables has dimension $j + 1$ and then use Proposition 2.3.1.   □

For arbitrary webs, without any further restriction on the relative position of the tangent spaces of the leaves at the origin besides pairwise transversality, it is not possible to improve the bound beyond the specialization of the above to $n = 2$. That is for an arbitrary quasi-smooth k-web

$$\ell^j(\mathcal{W}) \geq \min(k, j+1). \tag{2.9}$$

Remark 2.3.5. Recently Cavalier and Lehmann have drawn special attention to k-webs on $(\mathbb{C}^n, 0)$ for which the upper bounds (2.8) are sharp, see [27]. These have been labeled by them **ordinary webs**.

2.3.2 Bounds for the Rank

There is a **natural decreasing filtration** $F^\bullet \mathcal{A}(\mathcal{W})$ on the vector space $\mathcal{A}(\mathcal{W})$. The first term is, of course, $F^0 \mathcal{A}(\mathcal{W}) = \mathcal{A}(\mathcal{W})$ and for $j \geq 0$ the j-th piece of the filtration is defined as

$$F^j \mathcal{A}(\mathcal{W}) = \ker \left\{ \mathcal{A}(\mathcal{W}) \longrightarrow \left(\frac{\Omega^1(\mathbb{C}^n, 0)}{\mathfrak{m}^j \cdot \Omega^1(\mathbb{C}^n, 0)} \right)^k \right\},$$

with $\mathfrak{m}$ being the maximal ideal of $\mathcal{O}(\mathbb{C}^n, 0)$.

Lemma 2.3.6. *If* $\mathcal{W}$ *is a germ of quasi-smooth k-web on* $(\mathbb{C}^n, 0)$, *then*

$$\dim \frac{F^j \mathcal{A}(\mathcal{W})}{F^{j+1} \mathcal{A}(\mathcal{W})} \leq \max \left(0, k - \ell^{j+1}(\mathcal{W}) \right).$$

Proof. Let, as above, $h_1, \ldots, h_k$ be the linear parts at the origin of the submersions defining $\mathcal{W}$. Consider the linear map

$$\varphi : \quad \mathbb{C}^k \longrightarrow \mathbb{C}_{j+1}[x_1, \ldots, x_n]$$

$$(c_i)_{i=1}^k \longmapsto \sum_{i=1}^k c_i (h_i)^{j+1}.$$

From the definition of the space $\mathcal{L}^j(\mathcal{W})$, it is clear that the image of φ coincides with it. In particular,

$$\dim \ker(\varphi) = \max(0, k - \ell^{j+1}(\mathcal{W})).$$

If $(\eta_1, \ldots, \eta_k)$ is an abelian relation in $F^j \mathcal{A}(\mathcal{W})$, then for suitable complex numbers $\mu_1, \ldots, \mu_k$, the following identity holds true

$$(\eta_1, \ldots, \eta_k) = (\mu_1 (h_1)^j dh_1, \ldots, \mu_k (h_k)^j dh_k) \mod F^{j+1} \mathcal{A}(\mathcal{W}).$$

Consider the linear map taking $(\eta_1, \ldots, \eta_k) \in F^j \mathcal{A}(\mathcal{W})$ to the k-uple of complex numbers $(\mu_1, \ldots, \mu_k) \in \mathbb{C}^k$. Since $\sum \eta_i = 0$ it follows that this map induces an injection of $F^j \mathcal{A}(\mathcal{W})/F^{j+1} \mathcal{A}(\mathcal{W})$ into the kernel of φ. The lemma follows. $\square$

Corollary 2.3.7. *If $\mathcal{W}$ is a quasi-smooth k-web, then for $j \geq k - 2$*

$$F^j \mathcal{A}(\mathcal{W}) = 0 \,.$$

Proof. For $j \geq k - 2$, Eq. (2.9) reads as $\ell^{j+1}(\mathcal{W}) = k$. Therefore Lemma 2.3.6 implies

$$F^j \mathcal{A}(\mathcal{W}) = F^{j+1} \mathcal{A}(\mathcal{W}) \,.$$

Thus an element of $F^{k-2} \mathcal{A}(\mathcal{W})$ has a zero of infinite order at the origin. Since it is a k-uple of holomorphic 1-forms it has to be identically zero.   □

With what has been done so far, Bol's, respectively Chern's, bound for the rank of smooth k-webs on $(\mathbb{C}^2, 0)$, respectively on $(\mathbb{C}^n, 0)$, can be easily proved.

Theorem 2.3.8. *If $\mathcal{W}$ is a germ of quasi-smooth k-web on $(\mathbb{C}^n, 0)$, then*

$$\mathrm{rank}(\mathcal{W}) \leq \sum_{j=0}^{k-3} \max(0, k - \ell^{j+1}(\mathcal{W})) \,.$$

Moreover, if $\mathcal{W}$ is smooth, then

$$\mathrm{rank}(\mathcal{W}) \leq \pi(n,k) = \sum_{j=0}^{k-3} \max\left(0, k - (j+1)(n-1) - 1\right) \,.$$

Proof. It follows from the corollary above that $\mathcal{A}(\mathcal{W})$ is isomorphic as a vector space to

$$\bigoplus_{j=0}^{k-3} \frac{F^j \mathcal{A}(\mathcal{W})}{F^{j+1} \mathcal{A}(\mathcal{W})} \,.$$

If $\mathcal{W}$ is quasi-smooth the result follows promptly from Lemma 2.3.6. If moreover $\mathcal{W}$ is smooth, one can invoke Proposition 2.3.1 to conclude.   □

The number $\pi(n,k)$ appearing in the bound for the rank of smooth webs is **Castelnuovo's number**. It is the bound for the arithmetical genus of non-degenerate irreducible curves in $\mathbb{P}^n$ according to a classical result by Castelnuovo. In Chap. 3 Castelnuovo result will be recovered from Theorem 2.3.8 combined with Abel's addition Theorem.

Remark 2.3.9. Following [68], let $m = \lfloor \frac{k-1}{n-1} \rfloor$ and ϵ be the remainder of the division of $k - 1$ by $n - 1$. Thus $k - 1 = m(n-1) + \epsilon$ with $0 \leq \epsilon \leq n - 2$. Using this notation Castelnuovo's numbers can be expressed as

$$\pi(n,k) = \binom{m}{2}(n-1) + m\epsilon.$$

In this way one obtains a family of closed formulas for the bound of the rank of a k-web on $(\mathbb{C}^n, 0)$ according to the residue ϵ of $k-1$ modulo $n-1$.

Remark 2.3.10. Alternatively, one can set $\rho = \left\lfloor \frac{k-n-1}{n-1} \right\rfloor$ and ϵ equal to the remainder of the division of $k-n-1$ by $n-1$. Hence $k-n-1 = \rho(n-1)+\epsilon$ with $0 \le \epsilon \le n-2$. Castelnuovo's numbers admit the following alternative presentation

$$\pi(n,k) = (\epsilon + 1)\binom{\rho+2}{2} + (n-2-\epsilon)\binom{\rho+1}{2}.$$

The two distinct presentations are given here because the former is the usual one found in the literature, while the latter seems to be better adapted to some geometric constructions that will be carried out in Sect. 4.3.4 of Chap. 4.

Notice that for smooth webs the bound for the rank is attained if and only if the partial bounds provided by the combination of Proposition 2.3.1 with Lemma 2.3.6 are also attained. For later use, this remark is stated below as a corollary.

Corollary 2.3.11. *Let $\mathcal{W}$ be a germ of smooth k-web on $(\mathbb{C}^n, 0)$. If rank$(\mathcal{W}) = \pi(n,k)$, then*

$$\dim \frac{F^j \mathcal{A}(\mathcal{W})}{F^{j+1}\mathcal{A}(\mathcal{W})} = \max\left(0, k-(j+1)(n-1)-1\right)$$

for every $j \ge 0$.

2.3.3 Webs of Maximal Rank

One of the central problems in web geometry, and the central theme of this book, is the characterization of germs of smooth k-webs on $(\mathbb{C}^n, 0)$ for which rank$(\mathcal{W}) = \pi(n,k)$. They are called **webs of maximal rank**.

Theorem 2.3.8 recovers, and generalizes to arbitrary planar webs, the bound provided by Proposition 2.2.3 for germs of planar quasi-parallel webs. In particular the planar parallel webs are examples of webs of maximal rank, see Example 2.2.1. The 5-webs mentioned in Example 2.2.4 are also of maximal rank.

In dimension greater than two, webs of maximal rank are harder to come by. In contrast with the planar case, not every parallel web is of maximal rank. In the next section the parallel webs of maximal rank will be characterized in Proposition 2.4.3 and constraints on the distribution of conormals of maximal rank webs will be established.

2.4   Conormals of Webs of Maximal Rank

If $\mathcal{W}$ is a germ of smooth k-web of maximal rank, then the lower bounds for $\ell^j(\mathcal{W})$ given by Proposition 2.3.1 are attained, that is the equality

$$\ell^j(\mathcal{W}) = \min\left(k, j(n-1) + 1\right)$$

holds true for every positive integer j.

When the ambient space has dimension $n = 2$, these equalities do not impose any restriction on the web as Corollary 2.3.4 testifies. When n is at least three then the equalities above impose rather strong restrictions on the distributions of conormals of the web $\mathcal{W}$. Indeed, in the next few pages, the corresponding equality for $j = 2$—that is, $\ell^2(\mathcal{W}) = \min(k, 2(n-1) + 1)$—will be exploited and the following proposition will be proved.

Proposition 2.4.1. *Let $\mathcal{W}$ be a germ of smooth k-web on $(\mathbb{C}^n, 0)$. Suppose that $n \geq 3$ and $k \geq 2n + 1$. If $\ell^2(\mathcal{W}) = 2n - 1$, then there exists a non-degenerate rational normal curve Γ in $\mathbb{P}(T_0^*(\mathbb{C}^n, 0))$ containing the conormals of the web $\mathcal{W}$ at the origin.*

A Particular Case

For the sake of clarity, the case $n = 3$ of Proposition 2.4.1 will be presented here. It is harmless to assume that $k = 2n = 6$ even if the hypothesis for $n = 3$ reads $k \geq 7$.

Let $h_1, \ldots, h_6 \in \mathbb{C}_1[x_1, x_2, x_3]$ be six linear forms in general position and let $\mathcal{L}^2$ be the vector space contained in $\mathbb{C}_2[x_1, x_2, x_3]$ generated by theirs squares.

If $\dim \mathcal{L}^2 = 5$, since $\mathbb{C}_2[x_1, x_2, x_3]$ has dimension six, there is a hyperplane H through $0 \in \mathbb{C}_2[x_1, x_2, x_3]$ containing $\mathcal{L}^2$. If one now interprets the linear forms h_i as points in $\mathbb{P}^2 = \mathbb{P}\mathbb{C}_1[x_1, x_2, x_3]$ and considers the Veronese embedding of this $\mathbb{P}^2$ into $\mathbb{P}^5 = \mathbb{P}\mathbb{C}_2[x_1, x_2, x_3]$, as was done in the proof of Lemma 2.3.2, then the pull-back of $[H]$ to $\mathbb{P}^2$ is a conic containing $[h_1], \ldots, [h_6]$. This is the sought rational normal curve.

Dimension Shift and Reduction to Castelnuovo Lemma

As suggested by its statement, all the action in the proof of Proposition 2.4.1 will take place in $\mathbb{P}T_0^*(\mathbb{C}^n, 0) = \mathbb{P}^{n-1}$. To avoid carrying over a -1 throughout, instead of working with a k-web on $(\mathbb{C}^n, 0)$, it is convenient to consider a k-web on $(\mathbb{C}^{n+1}, 0)$. Of course with this shift on the dimension the hypotheses of Proposition 2.4.1 now read as

$$k \geq 2n + 3 \qquad \text{and} \qquad \ell^2(\mathcal{W}) = 2n + 1.$$

If $h_1, \ldots, h_k \in \mathbb{C}[x_0, \ldots, x_n]$ are the linear forms defining the tangent spaces of the leaves of $\mathcal{W}$ through the origin, then according to Lemma 2.3.2 the number of conditions imposed on hyperquadrics of $\mathbb{P}^n$ by the corresponding set of points $\mathcal{P} = \{[h_i] \mid i \in \underline{k}\} \subset \mathbb{P}^n$ is exactly $\ell^2(\mathcal{W})$. Therefore Proposition 2.4.1 is equivalent to the famous Castelnuovo Lemma.

Proposition 2.4.2 (Castelnuovo Lemma). *Let $\mathcal{P} \subset \mathbb{P}^n$ be a set of k points in general position. Suppose that $n \geq 2$ and $k \geq 2n + 3$. If $\mathcal{P}$ imposes only $2n + 1$ conditions on the linear system of quadric hypersurfaces $|\mathcal{O}_{\mathbb{P}^n}(2)|$, then $\mathcal{P}$ is contained in a rational normal curve Γ of degree n.*

Before dealing with the proof of Proposition 2.4.2 itself, which will follow [68, Chapter III], some basic properties of rational normal curves will be reviewed.

2.4.1 Rational Normal Curves

The rational normal curves in a projective space $\mathbb{P}^n$ are the ones that admit a parametrization of the form

$$\varphi: \quad \mathbb{P}^1 \longrightarrow \quad \mathbb{P}^n$$

$$[s:t] \longmapsto \left[a_0(s:t) : \cdots : a_n(s:t)\right]$$

where $a_0, \ldots, a_n$ form a basis of the space $\mathbb{C}_n[s, t]$ of binary forms of degree n. In other words, a rational normal curve Γ in $\mathbb{P}^n$ is the image of an embedding of $\mathbb{P}^1$ into $\mathbb{P}^n$ given by the complete linear system $|\mathcal{O}_{\mathbb{P}^1}(n)|$.

Clearly, the intersection of such a curve Γ with a hyperplane H consists of at most n points. If the hyperplane is generic, then the intersection has exactly n points, that is Γ has degree n. It turns out that this is the minimal degree among the non-degenerated curves in $\mathbb{P}^n$. Moreover the rational normal curve is the unique irreducible non-degenerated curve of degree n in $\mathbb{P}^n$, see Proposition 2.4.11 below.

Notice that any k distinct points on a rational normal curve $\Gamma \subset \mathbb{P}^n$ are automatically in general position with respect to the linear system of hyperplanes. Indeed, if a subset of cardinality $a \leq n$ is contained in a $\mathbb{P}^{a-2}$, then by choosing other $n - a + 1$ points and considering a hyperplane containing all these $n + 1$ points one arrives at a contradiction since a rational normal curve Γ intersects every hyperplane in at most deg $\Gamma = n$ points.

Parallel webs defined by points on a rational normal curve are the simplest examples of webs of maximal rank on $(\mathbb{C}^n, 0)$, with $n \geq 3$. More precisely,

Proposition 2.4.3. *Let $n \geq 2$ and $k \geq 2n + 3$ be integers. Let also $h_1, \ldots, h_k \in \mathbb{C}_1[x_0, x_1, \ldots, x_n]$ be pairwise distinct linear forms and $\mathcal{W} = \mathcal{W}(h_1, \ldots, h_k)$ be*

the corresponding parallel k-web on $(\mathbb{C}^{n+1}, 0)$. *Then* $\mathcal{P} = \{[h_1], \ldots, [h_k]\}$, *the corresponding set of points of* $\mathbb{P}^n$, *lies in a rational normal curve* Γ *of degree n if and only if* $\mathcal{W}$ *is smooth and of maximal rank.*

Proof. If $\mathcal{W}$ is smooth and of maximal rank, then Proposition 2.4.1 implies the result.

Reciprocally, if $\mathcal{P}$ is contained in a rational normal curve of degree n, then $\mathcal{W}$ is smooth because the points $[h_i]$ are in general position, see the discussion preceding the statement of the Proposition.

To prove that $\mathcal{W}$ is of maximal rank notice that the kernel of the restriction map

$$H^0\big(\mathbb{P}^n, \mathcal{O}_{\mathbb{P}^n}(j)\big) \longrightarrow H^0\big(\Gamma, \mathcal{O}_\Gamma(j)\big) \simeq H^0\big(\mathbb{P}^1, \mathcal{O}_{\mathbb{P}^1}(jn)\big)$$

has codimension at least $h^0(\mathbb{P}^1, \mathcal{O}_{\mathbb{P}^1}(jn)) = jn + 1$. Therefore, by Lemma 2.3.2,

$$\ell^j(\mathcal{W}) = \dim\big(\mathbb{C}(h_1)^j + \cdots + \mathbb{C}(h_k)^j\big) \leq jn + 1.$$

Hence Proposition 2.3.1 implies $\ell^j(\mathcal{W}) = \min(k, jn + 1)$.

Because $\mathcal{W}$ is a parallel web, $F^j \mathcal{A}(\mathcal{W})/F^{j+1}\mathcal{A}(\mathcal{W})$ not just embeds into the kernel of the map $\mathbb{C}^k \to \mathcal{L}^j(\mathcal{W})$ considered in the proof of Lemma 2.3.6, but is indeed isomorphic to it. Therefore

$$\mathrm{rank}(\mathcal{W}) = \sum_{j=0}^{k-3} \max\big(0, k - (j+1)n - 1\big)$$

as requested. $\qquad\qquad\qquad\qquad\qquad\qquad\qquad\qquad\qquad\qquad\qquad\qquad\qquad\quad\square$

Steiner's Synthetic Construction

The rational normal curves admit a nice geometric description: the so-called **Steiner construction**. Let $p_1, \ldots, p_{n+3} \in \mathbb{P}^n$ be $n + 3$ points in general position. For each i ranging from 1 to n, let Π_i be the $\mathbb{P}^{n-2}$ spanned by $p_1, \ldots, \widehat{p}_i, \ldots, p_n$. The hyperplanes containing Π_i form a family $H_i(s : t)$ with $[s : t] \in \mathbb{P}^1$. One can choose the parametrizations $[s : t] \mapsto H_i(s : t)$ in order to have

$$p_{n+1} = \bigcap_{i=1}^n H_i(0 : 1), \quad p_{n+2} = \bigcap_{i=1}^n H_i(1 : 0) \quad \text{and} \quad p_{n+3} = \bigcap_{i=1}^n H_i(1 : 1).$$

Proposition 2.4.4. *The set*

$$\Gamma = \bigcup_{[s:t] \in \mathbb{P}^1} \left(\bigcap_{i=1}^n H_i(s : t)\right)$$

is the unique rational normal curve passing through the points $p_1, \ldots, p_{n+3}$.

Proof. Because the points are in general position the expression between parenthesis defines for each $[s : t] \in \mathbb{P}^1$ a unique point of $\mathbb{P}^n$. Consequently Γ is a curve parametrized by $\mathbb{P}^1$.

Clearly it contains p_{n+1}, p_{n+2} and p_{n+3}. To see that it contains $p_1, \ldots, p_n$ notice that $p_i \in H_j(s : t)$ for every $[s : t] \in \mathbb{P}^1$ when $j \neq i$ and there exists $[s_i : t_i] \in \mathbb{P}^1$ such that $p_i \in H_i(s_i : t_i)$. It remains to show that the linear system defining Γ is $|\mathcal{O}_{\mathbb{P}^1}(n)|$.

Using an automorphism of $\mathbb{P}^n$ the points $p_1, \ldots, p_{n+1}$ can normalized as

$$ p_i = [0 : \ldots : 0 : \underbrace{1}_{\text{i-th entry}} : 0 : \ldots : 0] \quad i \in \underline{n+1}. $$

If $p_{n+2} = [a_0 : a_1 : \ldots : a_n]$ and $p_{n+3} = [b_0 : b_1 : \ldots : b_n]$, then it is a simple matter to verify that

$$ \mathbb{P}^1 \longrightarrow \mathbb{P}^n $$

$$ [s : t] \longmapsto \left[(a_1^{-1}s - b_1^{-1}t)^{-1} : \cdots : (a_{n+1}^{-1}s - b_{n+1}^{-1}t)^{-1} \right] $$

is a parametrization of Γ in the normalization above. Multiplying all the entries by $\prod_i (a_i^{-1}s - b_i^{-1}t)$ one ends up with $n + 1$ binary forms of degree n. Since $p_1, \ldots, p_{n+3}$ are not contained in any hyperplane these must generate the space of binary forms of degree n. □

2.4.2   Proof of Castelnuovo Lemma

A variant of the synthetic construction presented above allows to construct rank 3 quadrics—these are quadrics which in a suitable system of homogeneous coordinates are cut out by a polynomial of the form $x_0^2 + x_1^2 + x_2^2$—containing Γ. This construction will be the key to prove Proposition 2.4.2.

Let $\mathcal{P} = \{p_1, \ldots, p_{2n+2}, p_{2n+3}\}$ be a set of $2n + 3$ distinct points of $\mathbb{P}^n$ in general position, and $\Lambda \simeq \mathbb{P}^{n-2}$ be the linear span of $p_1, \ldots, p_{n-1}$.

Lemma 2.4.5. *If $\mathcal{P}$ imposes at most $2n + 1$ independent conditions on the space of quadrics, then there are at least $n - 1$ linearly independent quadrics containing $\Lambda \cup \mathcal{P}$.*

Proof. Let F_0 be a linear form (unique up to multiplication by an element of $\mathbb{C}^*$) vanishing on the span of $p_1, \ldots, p_n$ and G_0 be another one vanishing at $p_1, \ldots, p_{n-1}, p_{n+1}$. Any quadric containing Λ is cut out by an equation of the form $F_0 G - G_0 F$ for suitable linear forms $F, G \in \mathbb{C}_1[x_0, \ldots, x_n]$. Such a pair (F, G) is not unique, but is well defined modulo the addition of a multiple of (G_0, F_0). Hence the vector space of quadratic homogeneous forms vanishing on Λ has dimension $2n + 1$.

Further imposing that the quadrics contain the $n + 2$ points $p_n, p_{n+1}, \ldots, p_{2n+1}$ one sees that there are at least $2n + 1 - (n + 2)$, that is $n - 1$, linearly independent quadrics containing $\Lambda \cup \{p_1, \ldots, p_{2n+1}\}$. By hypothesis the space of quadrics containing $\{p_1, \ldots, p_{2n+1}\}$ coincides with the space of quadrics containing $\mathcal{P}$. ☐

Keeping the notation from the lemma above one can write down the $n - 1$ linearly independent quadrics $Q_1, \ldots, Q_{n-1}$ containing $\Lambda \cup \mathcal{P}$ in the form

$$Q_i = \det \begin{pmatrix} F_0 & F_i \\ G_0 & G_i \end{pmatrix} = F_0 G_i - G_0 F_i$$

for suitable linear forms $F_i, G_i \in \mathbb{C}_1[x_0, \ldots, x_n]$, $i \in \underline{n-1}$.

It is possible to recover a rational normal curve Γ from the quadrics just constructed. It will turn out that the variety X defined through the determinantal formula below

$$X = \left\{ p \in \mathbb{P}^n \;\middle|\; \mathrm{rank} \begin{pmatrix} F_0(p) & \cdots & F_{n-1}(p) \\ G_0(p) & \cdots & G_{n-1}(p) \end{pmatrix} \leq 1 \right\} \tag{2.10}$$

is a rational normal curve. To prove it, a couple of preliminary results are needed.

Lemma 2.4.6. *For any pair of complex numbers $(\lambda, \mu) \in \mathbb{C}^2$ distinct from $(0, 0)$ the linear forms $\lambda F_i + \mu G_i$, $i = 0, \ldots, n - 1$ are linearly independent.*

Proof. Because the quadrics $Q_1, \ldots, Q_{n-1}$ are linearly independent for any $\alpha = (\alpha_1, \ldots, \alpha_{n-1}) \in \mathbb{C}^{n-1} \setminus \{0\}$ the quadric $Q_\alpha = \sum_{i=1}^{n-1} \alpha_i Q_i$ cut out by

$$\det \begin{pmatrix} F_0 & \sum\limits_{i=1}^{n-1} \alpha_i F_i \\ G_0 & \sum\limits_{i=1}^{n-1} \alpha_i G_i \end{pmatrix} \tag{2.11}$$

is non-zero and still contains $\mathcal{P}$. If the linear forms $\lambda F_i + \mu G_i$, $i = 0, \ldots, n - 1$ are linearly dependent, then there exists $(\alpha_0, \ldots, \alpha_n) \in \mathbb{C}^{n+1} \setminus \{0\}$ such that

$$\alpha_0 (\lambda F_0 + \mu G_0) = \sum_{i=1}^{n-1} \alpha_i (\lambda F_i + \mu G_i).$$

Consequently the 2×2 matrix appearing in Eq. (2.11) has rank 1. Thus the quadric Q_α has rank at most 2. Since $\mathcal{P}$ is not contained in the union of two hyperplanes, the lemma follows. ☐

Lemma 2.4.7. *The restriction of any linear combination of the linear forms $F_1, \ldots, F_{n-1}$ to Λ is non-zero. Consequently, it can be assumed that for every $i = 1, \ldots, n - 1$, the linear form F_i satisfies*

$$p_1 , \ldots , p_{i-1} , \widehat{p}_i , p_{i+1} , \ldots , p_{n-1} \in \{F_i = 0\} .$$

Proof. Since $F_0(p_n) = 0$, $G_0(p_n) \neq 0$ and the quadrics Q_i contain p_n, the linear form F_i must vanish on p_n for every $i \in \underline{n-1}$. If some linear combination of $F_1, \ldots, F_{n-1}$ vanishes on Λ, then it has to be a complex multiple of F_0 because the span of Λ and p_n is the hyperplane cut out by F_0. This contradicts the linear independence of $F_0, \ldots, F_{n-1}$ established in the previous lemma and proves the first claim. The second claim follows immediately from plain linear algebra. □

Castelnuovo Lemma follows from the proposition below.

Proposition 2.4.8. *The variety X is the unique rational normal curve through* $p_1, \ldots, p_k$.

Proof. If, for $i = 0, \ldots, n - 1$ and for $[s : t] \in \mathbb{P}^1$, $H_{n-i}(s : t)$ stands for the hyperplane $\{sF_i + tG_i = 0\}$ then X can be described as below

$$X = \bigcup_{[s:t] \in \mathbb{P}^1} \left(\bigcap_{i=1}^{n} H_i(s : t) \right).$$

This has exactly the same form as the presentation of a rational normal curve through Steiner's construction, see Proposition 2.4.4. Consequently, X is a rational normal curve.

By construction, when $l > n$, $Q_\alpha(p_l) = 0$ but $(F_0(p_l), G_0(p_l)) \neq (0, 0)$. Therefore

$$\det \begin{pmatrix} F_i(p_l) & F_j(p_l) \\ G_i(p_l) & G_j(p_l) \end{pmatrix} = 0$$

for every pair i, j and every $l > n$. Thus X contains the points $p_{n+1}, \ldots, p_k$.

The careful reader has probably noticed that the inequality $k \geq 2n + 3$ has not been used so far. Only the weaker $k \geq 2n + 1$ played a role. To prove that $p_1, \ldots, p_n$ belong to X the stronger inequality now enters the stage. Observe that the quadric $Q_{ij} = F_i G_j - F_j G_i$ contains the $k - 2 \geq 2n + 1$ points $\mathcal{P} - \{p_i, p_j\}$. Remark 2.3.3 implies that these points impose at least $2n + 1$ conditions on the space quadrics. But, by hypothesis, the same holds true for $\mathcal{P}$. Thus Q_{ij} also contains p_i and p_j.
□

2.4.3 Normal Forms for Webs of Maximal Rank

For a quasi-smooth k-web $\mathcal{W} = \mathcal{W}(\omega_1, \ldots, \omega_k)$ on $(\mathbb{C}^n, 0)$ it is natural to consider $\ell^j(\mathcal{W})$ not just as an integer but as a germ of integer-valued function defined on $(\mathbb{C}^n, 0)$. The value at $x \in (\mathbb{C}^n, 0)$ is given by the dimension of the span of $\{\omega_i(x)^j ; i \in \underline{k}\}$ in the complex vector space $\mathrm{Sym}^j \Omega^1_x(\mathbb{C}^n, 0)$.

A priori this function does not need to be continuous but just lower semi-continuous. Nevertheless, as the reader can easily verify, when $\mathcal{W}$ is smooth and $F^{j-1}\mathcal{A}(\mathcal{W})/F^j \mathcal{A}(\mathcal{W})$ has maximal dimension then $\ell^j(\mathcal{W})$ is constant.

To make further references easier, this fact is stated below as a lemma.

Lemma 2.4.9. *Let $\mathcal{W}$ be a smooth k-web. If*

$$\dim \frac{F^{j-1}\mathcal{A}(\mathcal{W})}{F^j \mathcal{A}(\mathcal{W})} = \min\left(0, k - j(n-1) - 1\right)$$

then the integer-valued function $\ell^j(\mathcal{W}) : (\mathbb{C}^n, 0) \to \mathbb{N}$ is constant and equal to $j(n-1) + 1$.

Combined with Castelnuovo Lemma, or rather with Proposition 2.4.1, the Lemma above yields the following normal form for webs of maximal rank up to second order.

Proposition 2.4.10. *Let $\mathcal{W} = \mathcal{F}_1 \boxtimes \cdots \boxtimes \mathcal{F}_k$ be a germ of smooth k-web on $(\mathbb{C}^n, 0)$. Suppose that $n \geq 3$ and $k \geq 2n + 1$. If*

$$\dim \frac{F^1\mathcal{A}(\mathcal{W})}{F^2\mathcal{A}(\mathcal{W})} = k - 2n + 1$$

then there exist a coframe $\varpi = (\varpi_0, \ldots, \varpi_{n-1})$ on $(\mathbb{C}^n, 0)$ and k germs of holomorphic functions $\theta_1, \ldots, \theta_k \in \mathcal{O}(\mathbb{C}^n, 0)$ such that, for every $i \in \underline{k}$, one has:

$$\mathcal{F}_i = \left[\sum_{q=0}^{n-1} (\theta_i)^q \varpi_q \right].$$

Proof. According to Lemma 2.4.9 the function $\ell^2(\mathcal{W})$ is constant and equal to $2(n-1) + 1$. Proposition 2.4.1 implies the existence, for every $x \in (\mathbb{C}^n, 0)$, of a rational normal curve $\Gamma_x \subset \mathbb{P}T^*_x(\mathbb{C}^n, 0)$ of degree $n - 1$ containing the conormals of the defining foliations of $\mathcal{W}$.

Therefore it is possible to choose holomorphic 1-forms $\varpi_0, \ldots, \varpi_{n-1} \in \Omega^1(\mathbb{C}^n, 0)$ such that, for every $x \in (\mathbb{C}^n, 0)$, the rational normal curve Γ_x is parameterized by

$$t \longmapsto \sum_{q=0}^{n-1} t^q \varpi_q(x).$$

This parametrization can be chosen in such a way that none of the foliations $\mathcal{F}_i$ have conormal corresponding to $t = \infty$. Thus, for every $i \in \underline{k}$, the foliation $\mathcal{F}_i$ will be induced by $\sum_{q=0}^{n-1} (\theta_i)^q \varpi_q$ for a suitable germ of holomorphic function θ_i on $(\mathbb{C}^n, 0)$. $\qquad \square$

2.4.4   A Generalization of Castelnuovo Lemma

Proposition 2.4.10 can be seen as the starting of the proof of the algebraization of webs of maximal rank to be presented in Chap. 5. As it has been made clear above, Proposition 2.4.10 is an easy consequence of Castelnuovo Lemma. Loosely phrased, Castelnuovo Lemma says that if sufficiently many points in general position impose the minimal number of conditions on the space of quadric hypersurfaces, then they must lie on particularly simple curves: the rational normal curves. A testimony of the *simplicity* of rational normal curves is the following proposition.

Proposition 2.4.11. *If C is a non-degenerate irreducible projective curve in $\mathbb{P}^n$, then $\deg C \geq n$. Moreover, if the equality holds, then C is a rational normal curve.*

Proof. If C is non-degenerate, then there exist n points on C that are in general position, otherwise C would be contained in a hyperplane. Intersecting C with the hyperplane H determined by n such points shows that the degree of C is at least n.

To prove the second part let $p_1, \ldots, p_{n-1}$ be $n-1$ general points of C and let Σ be the $\mathbb{P}^{n-2}$ spanned by them. By hypothesis each generic hyperplane containing Σ intersects C in exactly one point away from Σ. Therefore there is an injective map from the set of hyperplanes containing Σ, which is nothing else than a $\mathbb{P}^1$, to C. Thus C is rational. Therefore C is parametrized by $n+1$ homogenous binary forms of degree equal to $\deg C = n$. Since C is non-degenerate, these $n+1$ binary forms must generate the space of degree n binary forms. In other words, C is a rational normal curve of degree n. $\square$

It is natural to inquire what can be said about sufficiently many points imposing a close to minimal number of conditions on the space of hyperquadrics. For instance, one can ask if they lie on *simple* varieties.

Of course, to be more precise, the meaning of *simple varieties* must be spelled out. One possibility is to look for non-degenerate irreducible varieties of minimal degree. For that sake it is important to generalize Proposition 2.4.11 to irreducible non-degenerated varieties of $\mathbb{P}^n$ of arbitrary dimension. The first part of the statement generalizes promptly as shown below.

Proposition 2.4.12. *If X is a non-degenerate irreducible subvariety of $\mathbb{P}^n$, then $\deg X \geq \operatorname{codim} X + 1$.*

Proof. Take $m + 1 = \operatorname{codim}(X) + 1$ generic points on X. Because X is non-degenerate they span a $\mathbb{P}^m$ intersecting X in at least $m + 1$ points. To conclude, one has to verify that for a generic choice of $m + 1$ points there are no positive dimensional component in the corresponding intersection $\mathbb{P}^m \cap X$.

For that sake let k be the dimension of the intersection of X with a generic $\mathbb{P}^m$. If $p_1, \ldots, p_{m-k+1} \in X$ are $m - k + 1$ generic points, then their linear span $\Sigma \simeq \mathbb{P}^{m-k}$ intersects X in a finite number of points. If Λ is the set of all the projective spaces $\mathbb{P}^{m-k+1}$ contained in $\mathbb{P}^n$ and containing Σ, then $\Lambda \simeq \mathbb{P}^{n-m+k}$.

On the one hand $\dim X = n - m$, while on the other hand

$$X - \Sigma = \bigcup_{\mathbb{P}^{n-m+k} \in \Lambda} (\mathbb{P}^{m-k+1} - \Sigma) \cap X .$$

implies that $\dim X = n - m + k$. Thus $k = 0$, that is, X intersects a generic $\mathbb{P}^m$ in a finite number of points. $\qquad\square$

The second part also does generalize but the generalization, which can be traced back at least to Bertini, is by no means evident.

Theorem 2.4.13. *If V is an irreducible non-degenerate projective subvariety of $\mathbb{P}^n$ with $\deg V = \operatorname{codim} V + 1$, then*

1. *V is $\mathbb{P}^n$; or*
2. *V is a rational normal scroll; or*
3. *V is a cone over the Veronese surface $v_2(\mathbb{P}^2) \subset \mathbb{P}^5$; or*
4. *V is a hyperquadric.*

The proof of this theorem would take the exposition too far afield, and therefore will not be presented here. For a modern reference, see [47] for instance.

A **rational normal scroll** of dimension m in $\mathbb{P}^n$ is characterized, up to an automorphism of $\mathbb{P}^n$, by m non-negative integers $a_1, \ldots, a_m$ summing up to $n - m + 1$ and can be described as follows. Decompose $\mathbb{C}^{n+1}$ as a direct sum $\oplus \mathbb{C}^{a_i+1}$ and consider parametrizations $\varphi_i : \mathbb{P}^1 \to \mathbb{P}^{a_i} \subset \mathbb{P}^n$ of rational normal curves in the corresponding projective subspaces.[2] If $\Sigma(p)$ is the $\mathbb{P}^{m-1}$ spanned by $\varphi_1(p), \ldots, \varphi_m(p)$, then the associated rational normal scroll is

$$S_{a_1, \ldots, a_m} = \bigcup_{p \in \mathbb{P}^1} \Sigma(p) .$$

Notice that the rational normal curves are rational normal scrolls, with $m = 1$ according to the definition above.

It is natural to consider the rational normal scrolls as higher-dimensional analogues of rational normal curves. The analogies between rational normal curves and scrolls do not amount only to similar definitions and to the fact that both are varieties of minimal degree. They encompass many other aspects. For instance, the rational normal scrolls of dimension m in $\mathbb{P}^n$ admit determinantal presentations, similar to the presentation (2.10) used for rational normal curves. More precisely, if $F_0, \ldots, F_{n-m}, G_0, \ldots, G_{n-m}$ are linear forms such that for any pair of complex numbers $(\lambda, \mu) \neq (0, 0)$, the linear forms $\{\lambda F_i + \mu G_i\}_{i=0, \ldots, n-m}$ are linearly independent (compare with Lemma 2.4.6) then

$$X = \left\{ \operatorname{rank} \begin{pmatrix} F_0 & \cdots & F_{n-m} \\ G_0 & \cdots & G_{n-m} \end{pmatrix} \leq 1 \right\}$$

---

[2] When $a_i = 0$, the linear subspace $\mathbb{P}^{a_i}$ is nothing but a point $p_i \in \mathbb{P}^n$. In this case, the following convention is adopted: the *rational normal curve* in $\mathbb{P}^{a_i}$ is not a curve, but the point p_i.

is a rational normal scroll of dimension m. Moreover, any rational normal scroll can be presented in this way.

Another testimony of the similarity between rational normal curves and scrolls is the following generalization of Castelnuovo Lemma.

Proposition 2.4.14 (Generalized Castelnuovo Lemma). *Let $\mathcal{P} \subset \mathbb{P}^n$ be a set of k points in general position. Suppose that $n \geq 2$ and $k \geq 2n + 1 + 2m$. If $\mathcal{P}$ imposes $2n + m$ conditions on the linear system of quadric hypersurfaces, then $\mathcal{P}$ is contained in a rational normal scroll of dimension m.*

The proof of Castelnuovo Lemma presented in Sect. 2.4.2 is the specialization to $m = 1$ of the Eisenbud–Harris proof of the generalized Castelnuovo Lemma. Having at hand the determinantal presentation of a rational normal scroll given above the reader should not have difficulties to recover the original proof as found in [68, pages 103–106].

While Castelnuovo Lemma is essential to the proof of Trépreau's algebraization theorem, to be carried out in Chap. 5, the implications of the generalized Castelnuovo Lemma to web geometry, if any, remain to be unfolded.

Chapter 3
Abel's Addition Theorem

So far, not many examples of abelian relations for webs appeared in this book. Besides the abelian relations for hexagonal 3-webs, the polynomial abelian relations for parallel webs (see Example 2.2.1), and the abelian relations for the planar quasi-parallel webs discussed in Example 2.2.4, which are by the way also polynomial, no other example was studied.

The main result of this chapter, Abel's addition Theorem, remedies this unpleasant state of affairs. It establishes, and is essentially equivalent to, an injection of the space of global abelian differentials—also known as Rosenlicht's or regular differentials—on a reduced projective curve C into the space of abelian relations of the dual web $\mathcal{W}_C$.

The exposition that follows renounces conciseness in favor of clarity. First the result is proved for smooth projective curves avoiding the technical difficulties inherent to the singular case. Only then the case of an arbitrary reduced projective curve is dealt with.

The readers familiar with Castelnuvo's bound for the arithmetical genus of irreducible non-degenerate projective curves will promptly realize that Castelnuovo curves (briefly described in Sect. 3.3) give rise to webs of maximal rank. The readers who are not, will get acquainted with Castelnuovo's bound since it can be seen as a joint corollary of the bound for the rank proved in Chap. 2 and of Abel's addition Theorem. While most of the arguments laid down in Chap. 2 to bound $\mathcal{A}(\mathcal{W})$ can be found in modern textbooks dealing with Castelnuovo Theory, the same cannot be said about the use of Abel's addition Theorem. At present, the proof of Castelnuovo's bound found in textbooks makes use of some basic results about the cohomology of projective varieties beside Castelnuovo Lemma.

Although not devoid of weaknesses when compared to the modern approach, the path to Castelnuovo's bound through Abel's addition Theorem has its own strong points. For instance, it makes it rather simple to obtain bounds for the genus of curves on other projective varieties, as explained in Sect. 3.4.

© Springer International Publishing Switzerland 2015
J.V. Pereira, L. Pirio, *An Invitation to Web Geometry*, IMPA Monographs 2,
DOI 10.1007/978-3-319-14562-4_3

3.1 Abel's Theorem I: Smooth Curves

3.1.1 Trace Under Ramified Coverings

Let X and Y be two smooth connected complex curves. A holomorphic map $f :$ $X \to Y$ is a **finite ramified covering** if it is surjective and proper. The **degree** of such a covering is defined as the cardinality of the pre-image of any of its regular values.

For X and Y as above, let $f : X \to Y$ be a finite ramified covering of degree k. For a regular value $q \in Y$ of f and a meromorphic 1-form ω defined at a neighborhood of $f^{-1}(q)$, let us define the **trace of ω relative to f at q** as the germ of meromorphic 1-form

$$\operatorname{tr}_{f,q}(\omega) = \sum_{i=1}^{k} g_i^*(\omega) \in \Omega^1(Y,q),$$

where $g_1, \ldots, g_k : (Y,q) \to X$ stand for the local inverses of f at q.

Proposition 3.1.1. *Let X and Y be two smooth, compact, and connected complex curves. If $f : X \to Y$ is a finite ramified covering, ω is a meromorphic 1-form globally defined on X and q is a regular value of f, then $\operatorname{tr}_{f,q}(\omega)$ extends to a unique meromorphic 1-form $\operatorname{tr}_f(\omega)$, which does not depend on q, and is globally defined on Y. Moreover, if ω is holomorphic, then $\operatorname{tr}_f(\omega)$ is also holomorphic.*

The meromorphic 1-form $\operatorname{tr}_f(\omega)$ globally defined on Y is, by definition, the **trace of ω relative to f**.

Proof of Proposition 3.1.1. For q varying among the regular values of f, the meromorphic 1-forms $\operatorname{tr}_{f,q}$ patch together to a meromorphic 1-form η defined on the whole complement of the set of critical values of f. Furthermore, if ω is holomorphic, then the same will be true for η.

Now, if $q \in Y$ is a critical value of f, then some point p in the fiber $f^{-1}(q)$ is a critical point. Although it is not possible to consider a local inverse $g : (Y,q) \to$ (X, p), the map $f : (X, p) \to (Y,q)$ is, in suitable local analytic coordinates x, y, the monomial map $f(x) = x^n = y$ for some positive integer n. Because X is compact, the set of critical values of f is finite. Therefore it suffices to consider the trace of the monomial maps from the disc $\mathbb{D}$ to itself to prove that η extends through the critical set of f.

For a point distinct from the origin, there are exactly n local inverses for $f(x) = x^n$: the distinct branches of $\sqrt[n]{x}$. One passes from one to another via multiplication by powers of ξ_n, a primitive n-th root of unity. More explicitly, if $\gamma(y) = y^{1/n}$ stands for a fixed local inverse of f (on any simply connected domain U in $\mathbb{C}^*$), then the local inverses of f on U are the functions $\xi_n^i \gamma$ for $i = 0, \ldots, n-1$.

Let us first consider the case of the trace by f of a monomial meromorphic differential $x^m dx$, with $m \in \mathbb{Z}$. On U, one has

$$\mathrm{tr}_f(x^m dx) = \sum_{i=0}^{n-1}(\xi_n^i\gamma)^*(x^m dx) = \sum_{i=0}^{n-1}(\xi_n^i y^{1/n})^m d(\xi_n^i y^{1/n})$$

$$= \Big[\sum_{i=0}^{n-1}\xi_n^{i(m+1)}\Big]y^{m/n}d(y^{1/n}) = \frac{1}{n}\Big[\sum_{i=0}^{n-1}\xi_n^{i(m+1)}\Big]y^{\frac{m+1}{n}-1}dy .$$

Consequently,

$$\mathrm{tr}_f(x^m dx) = \begin{cases} y^{\frac{m+1}{n}-1}dy & \text{if } m+1 = 0 \bmod n , \\ 0 & \text{otherwise} . \end{cases} \qquad (3.1)$$

Since the result does not depend on γ and because U is arbitrary, it follows that the trace of $x^m dx$ relative to f extends meromorphically at the origin.

The general case follows by first writing formally

$$\mathrm{tr}_f\Big(\Big(\sum_{k\geq m}c_k x^k\Big)dx\Big) = \sum_{k\geq m}c_k\mathrm{tr}_f(x^k dx)$$

and then by verifying that the right-hand side of this equality defines a meromorphic differential form at the origin if one assumes that $\sum_{k\geq m}c_k x^k$ is a meromorphic function of x (this is left as an exercise to the reader).

Finally, (3.1) shows that $\mathrm{tr}_f(x^k dx)$ is holomorphic at the origin if $k \geq 0$. The last statement of the proposition follows immediately. $\square$

Beware that there are meromorphic 1-forms with holomorphic trace as one can promptly infer from (3.1).

Remark 3.1.2. The algebraically inclined reader familiar with Kähler differentials and field extensions might prefer to define the trace relative to f as follows. Let $f : X \to Y$ be a finite ramified covering. Viewing X and Y as complex algebraic curves makes it natural to consider the induced finite field extension $f^* : \mathbb{C}(Y) \to \mathbb{C}(X)$ of the corresponding function fields. In this case, a rational function $\phi \in \mathbb{C}(X)$ has trace $\mathrm{tr}_f(\phi) \in \mathbb{C}(Y)$ equal to the trace of the endomorphism $\psi \mapsto \phi\psi$ of the finite dimensional $\mathbb{C}(Y)$-vector space $\mathbb{C}(X)$. Let $t \in \mathbb{C}(Y)$ be such that dt generates $\Omega_{\mathbb{C}(Y)}$ as a $\mathbb{C}(Y)$-module.[1] Therefore $f^*(dt) = df^*(t)$ generates the $\mathbb{C}(X)$-module of Kähler differentials on X. Hence any global meromorphic 1-form ω on X is written $\omega = \varphi\, df^*(t)$ for a certain meromorphic function $\varphi \in \mathbb{C}(X)$. The trace of ω relative to f is algebraically defined as $\frac{1}{[\mathbb{C}(X):\mathbb{C}(Y)]}\mathrm{tr}_f(\varphi)dt$.

---

[1]Here $\Omega_{\mathbb{C}(Y)}$ stands for the module of Kähler differentials over $\mathbb{C}(Y)$: by definition, it is the $\mathbb{C}(Y)$-module spanned by elements of the form $d\varphi$ for $\varphi \in \mathbb{C}(Y)$ subjected to the following relations: $dc = 0$ for every $c \in \mathbb{C}$; and $d(\phi\varphi) = \phi d\varphi + \varphi d\phi$ for every $\phi, \varphi \in \mathbb{C}(Y)$, see [70, II.§8].

3.1.2   Trace Relative to the Family of Hyperplanes

Let now C be a smooth and irreducible projective curve of degree k in $\mathbb{P}^n$, H_0 be a hyperplane intersecting C transversely, and ω be a meromorphic 1-form defined on a neighborhood of $H_0 \cap C$ in C. Let us consider the germs of holomorphic maps $p_1, \ldots, p_k : (\check{\mathbb{P}}^n, H_0) \to C$ such that, as zero-cycles on C, one has

$$H \cdot C = p_1(H) + \cdots + p_k(H)$$

for every $H \in (\check{\mathbb{P}}^n, H_0)$.

The **trace of** ω **at** H_0 **relative to the family of hyperplanes**, denoted by $\mathrm{Tr}_{H_0}(\omega)$, is defined through the formula

$$\mathrm{Tr}_{H_0}(\omega) = \sum_{i=1}^{k} p_i^*(\omega).$$

It is clearly a germ of meromorphic differential 1-form. As the trace relative to a ramified covering, it extends meromorphically to the whole projective space $\check{\mathbb{P}}^n$ as proved in the next section. This is essentially the content of Abel's addition Theorem.

3.1.3   Abel's Theorem for Smooth Curves

The next result is the version for smooth curves of what is called by web geometers Abel's addition Theorem, or just Abel's Theorem. The readers are warned that authors with other backgrounds might call a different, but essentially equivalent, statement by the same name. For a thorough discussion about the original version(s) of Abel's theorem, see [80].

Theorem 3.1.3 (Abel's Addition Theorem for Smooth Curves). *If ω is a meromorphic 1-form on a smooth projective curve $C \subset \mathbb{P}^n$, then the germ $\mathrm{Tr}_{H_0}(\omega)$ extends to a unique meromorphic 1-form $\mathrm{Tr}(\omega)$ globally defined on $\check{\mathbb{P}}^n$ which does not depend on H_0. Moreover, ω is a holomorphic 1-form on C if and only if $\mathrm{Tr}(\omega) = 0$.*

To prove Theorem 3.1.3, let $\check{U}_C \subset \check{\mathbb{P}}^n$ be the Zariski open subset formed by the hyperplanes $H \subset \mathbb{P}^n$ which intersect C at $k = \deg C$ distinct points. In other words, $\check{U}_C$ is the complement in $\check{\mathbb{P}}^n$ of the discriminant of the dual web $\mathcal{W}_C$ associated with C. The construction of $\mathrm{Tr}_{H_0}(\omega)$ made above can be made for any hyperplane $H \in \check{U}_C$. The results patch together to define a meromorphic 1-form $\mathrm{Tr}(\omega)$ on $\check{U}_C$.

To extend $\mathrm{Tr}(\omega)$ through the discriminant of $\mathcal{W}_C$, one will use a relation between the trace under a ramified covering and the trace relative to the family of hyperplanes.

To draw this relation, let ℓ be a line in $\check{\mathbb{P}}^n$. It corresponds to a pencil of hyperplanes in $\mathbb{P}^n$ whose base locus is equal to $\Pi = \check{\ell} \subset \mathbb{P}^n$, the $\mathbb{P}^{n-2}$ dual to ℓ. It will be convenient, although not strictly necessary, to assume that Π does not intersect C. Let us define

$$\pi_\ell : C \to \ell \simeq \mathbb{P}^1$$

as the morphism that associates with a point of C the hyperplane in ℓ containing it. Clearly it is a ramified covering, thus the trace $\mathrm{tr}_{\pi_\ell}(\omega)$ makes sense for any meromorphic 1-form ω on C.

Lemma 3.1.4. *For every meromorphic 1-form ω on C, the trace of ω under π_ℓ coincides with the pull-back to ℓ of the trace of ω relative to the family of hyperplanes, that is*

$$\mathrm{tr}_{\pi_\ell}(\omega) = i^* \mathrm{Tr}(\omega) \,,$$

where $i : \ell \hookrightarrow \check{\mathbb{P}}^n$ is the natural inclusion.

Proof. Consider the composition $\varphi = i \circ \pi_\ell : C \to \check{\mathbb{P}}^n$. The image of a point $q \in C$ is the hyperplane H containing both the point q and the linear space $\Pi = \check{\ell}$.

The hyperplane H intersects C at q, and at other $k-1$ points of C with multiplicities taken into account. All these other points are also mapped to H by φ. Thus, the functions $p_i \circ i : \ell \cap \check{U} \to C$ are local inverses of φ at any $H \in \check{U}_C$. Hence

$$\mathrm{tr}_{\pi_\ell}(\omega) = \mathrm{tr}_{i \circ \pi_\ell}(\omega) = \sum_{i=1}^{k} (p_i \circ i)^* \omega = i^* \mathrm{Tr}(\omega)$$

at the generic point of ℓ. The lemma follows. $\qquad\qquad\square$

Back to the proof of Abel's Theorem, recall that $\mathrm{Tr}(\omega)$ is defined all over the Zariski open set $\check{U}_C$. If it does not extend meromorphically to the whole $\check{\mathbb{P}}^n$, then its pull-back to a generic line $\ell \subset \check{\mathbb{P}}^n$ has an essential singularity at one of the points of $\ell \cap \Delta(\mathcal{W}_C)$. Lemma 3.1.4 implies the existence of an essential singularity for $\mathrm{tr}_{\pi_\ell}(\omega)$. But this cannot be the case according to Proposition 3.1.1.

Assume now that ω is holomorphic. Then for any generic line $\ell \subset \check{\mathbb{P}}^n$, the trace $\mathrm{tr}_{\pi_\ell}(\omega)$ is holomorphic, according to Proposition 3.1.1. Since there is no non-trivial holomorphic 1-form on $\ell \simeq \mathbb{P}^1$, this means that $\mathrm{tr}_{\pi_\ell}(\omega) = 0$. By Lemma 3.1.4, this implies that the restriction of $\mathrm{Tr}(\omega)$ on a generic line in $\check{\mathbb{P}}^n$ is trivial. This shows that $\mathrm{Tr}(\omega) = 0$ when ω is assumed to be holomorphic.

It remains to establish the converse implication. To prove it, let us suppose that ω is not holomorphic. If $x \in C$ is a pole of ω, then the generic hyperplane $H \subset \mathbb{P}^n$ through x intersects C transversely and avoids all the other poles of ω. Thus, in a neighborhood of H in $\check{\mathbb{P}}^n$, the trace of ω is the sum of the pull-back by a holomorphic map of a meromorphic, but not holomorphic, 1-form with other $\deg(C) - 1$ holomorphic 1-forms. Hence $\mathrm{Tr}(\omega)$ has a non-empty polar set and, in particular, is not zero.   □

3.1.4 Abelian Relations for Algebraic Webs

Theorem 3.1.3 can be interpreted in terms of webs and abelian relations instead of projective curves and holomorphic 1-forms. More precisely,

Theorem 3.1.5. *If C is a smooth projective curve of degree k and H_0 is a hyperplane intersecting it transversely, then the space of holomorphic 1-forms on C injects into the space of abelian relations of the dual web $\mathcal{W}_C(H_0)$.*

Proof. Let H_0 be a hyperplane intersecting C transversely in k points, and let $p_i : (\check{\mathbb{P}}^n, H_0) \to C$ be the germs of holomorphic functions such that $H \cdot C = \sum_i p_i(H)$ for all $H \in (\check{\mathbb{P}}^n, H_0)$. Recall from Chap. 1 that the k-web $\mathcal{W}_C(H_0)$ is defined by the submersions $p_1, \ldots, p_k$, that is $\mathcal{W}_C(H_0) = \mathcal{W}(p_1, \ldots, p_k)$.

If ω is a non-trivial holomorphic 1-form on C, then it is automatically closed, for dimensional reasons. Since the exterior differential commutes with pull-backs, the 1-forms $p_i^* \omega$ are also closed. Moreover for every i, the pull-back $p_i^* \omega$ defines the same foliation as the one defined by p_i, ie. $p_i^* \omega \wedge dp_i = 0$. Abel's addition theorem, in turn, implies that

$$\mathrm{Tr}(\omega) = p_1^*(\omega) + \cdots + p_k^*(\omega) = 0$$

holds identically on $(\check{\mathbb{P}}^n, H_0)$. Therefore $(p_1^* \omega, \ldots, p_k^* \omega)$ is an abelian relation of $\mathcal{W}_C(H_0)$.

It follows that the injective linear map

$$H^0\big(C, \Omega_C^1\big) \longrightarrow \Omega^1\big(\check{\mathbb{P}}^n, H_0\big)^k$$

$$\omega \longmapsto \big(p_1^* \omega, \ldots, p_k^* \omega\big)$$

factors through $\mathcal{A}(\mathcal{W}_C) \subset \Omega^1(\mathbb{P}^n, H_0)^k$.   □

Recall that $g(C)$—the **genus** of C—coincides with $h^0(C, \Omega_C^1)$, the dimension of the vector space $H^0(C, \Omega_C^1)$ of holomorphic 1-forms on C.

Corollary 3.1.6. *If C is a smooth projective curve, then for any hyperplane H_0 intersecting it transversely, one has*

$$\text{rank}\big(\mathcal{W}_C(H_0)\big) \geq g(C).$$

In Chap. 4, one will see that this lower bound is in fact an equality.

3.1.5 Castelnuovo's Bound

When read backwards Corollary 3.1.6 yields the famous Castelnuovo's bound on the genus of projective curves. More precisely,

Theorem 3.1.7 (Castelnuovo's Bound). *If C is a smooth connected non-degenerate projective curve in $\mathbb{P}^n$ of degree k, then*

$$g(C) \leq \pi(n,k).$$

Proof. For a generic H_0, the web $\mathcal{W}_C(H_0)$ is smooth according to Proposition 1.4.5. Chern's bound on the rank of smooth webs, see Theorem 2.3.8, combined with Corollary 3.1.6 implies the result.   □

It is instructive to compare this proof of Castelnuovo's bound with the usual textbook proof. The first step in both proofs relies on the bounds for the number of conditions imposed by points on the complete linear systems of hypersurfaces on the relevant projective space. While the former proof uses Abel's addition theorem to conclude, the latter appeals to Riemann–Roch Theorem instead. For a thorough discussion on this matter, see [34].

3.2 Abel's Theorem II: Arbitrary Curves

When studying germs of smooth algebraic webs $\mathcal{W}_C(H_0)$, it is hard to tell whether the curve C is smooth or not. At first sight such a web only exhibits properties of C valid at a neighborhood of the transversal hyperplane H_0. For smooth curves, we have just explained how the holomorphic differentials give rise to abelian relations for the dual web.

It is then natural to ask:

(a) are there another abelian relations for algebraic webs?
(b) which kind of *differentials* on a singular curve give rise to abelian relations for the dual web?

Question (a) will be treated in Chap. 4, while question (b) is the subject of the present section. Before dealing with it, some conventions are settled below.

Conventions on Singular Curves

Given a reduced analytic curve X (always assumed of finite type in what follows), its **desingularization** will be denoted by $v = v_X : \overline{X} \to X$.

A **meromorphic 1-form** on X is nothing else than a meromorphic 1-form ω on the smooth part of X such that $v^*\omega$, its pull-back to $\overline{X}$, extends to the whole $\overline{X}$ as a meromorphic 1-form. The sheaf of meromorphic differentials on a curve X, singular or not, will be denoted by $\mathcal{M}_X$.

For an arbitrary curve X and a smooth irreducible curve Y, a morphism $f : X \to Y$ will be called a **finite ramified covering**, if the restriction of $\overline{f} = f \circ v_X$ to each of the irreducible components of $\overline{X}$ is a finite ramified covering as defined in Sect. 3.1.1.

3.2.1 Residues and Traces

Let us assume that p is a smooth point of a curve X as above, and x is a local holomorphic coordinate on X centered at it. If ω is a germ of meromorphic differential at p, then

$$\omega = \sum_{i=i_0}^{\infty} a_i x^i dx$$

for some $i_0 \in \mathbb{Z}$ and suitable complex numbers a_i. The **residue of ω at p** is the complex number

$$\mathrm{Res}_p(\omega) = \begin{cases} a_{-1} & \text{if } i_0 \leq -1; \\ 0 & \text{otherwise.} \end{cases}$$

It is a simple matter to verify that this definition does not depend on the local coordinate x. One possibility is to notice that the residue can be determined through the integral formula

$$\mathrm{Res}_p(\omega) = \frac{1}{2i\pi} \int_\gamma \omega \tag{3.2}$$

for any sufficiently small (positively oriented) loop γ around p.

The following properties can be easily verified:

1. $\mathrm{Res}_p : \mathcal{M}_{X,p} \to \mathbb{C}$ is $\mathbb{C}$-linear;
2. $\mathrm{Res}_p(\omega) = 0$ when ω is holomorphic at p ;
3. $\mathrm{Res}_p(f^n df) = 0$ for all $f \in \mathcal{O}_{X,p}$ when $n \neq -1$;
4. $\mathrm{Res}_p(f^{-1} df) = v_p(f)$ for all meromorphic germs f at p (where v_p stands for the valuation associated with p).

Residues at Singular Points

Let us now assume that X is singular and let ω be a meromorphic differential 1-form on it. Let $v : \overline{X} \to X$ be the desingularization of X. The residue of ω at a singular point $p \in X_{\text{sing}}$ is defined as

$$\text{Res}_p(\omega) = \sum_{q \in v^{-1}(p)} \text{Res}_q(v^*(\omega)). \tag{3.3}$$

It is completely determined by the germ of ω at p.

Given a ramified covering $f : X \to Y$ between a possibly singular curve X and a smooth and irreducible curve Y, the trace relative to f of any meromorphic 1-form ω on X is defined by the relation

$$\text{tr}_f(\omega) = \text{tr}_{\overline{f}}(v_X^*\omega).$$

Proposition 3.2.1. *Let X and Y be curves with Y smooth and irreducible. If $f : X \to Y$ is a ramified covering then, for every $\omega \in H^0(X, \mathcal{M}_X)$ and every $p \in Y$, one has*

$$\text{Res}_p(\text{tr}_f(\omega)) = \sum_{q \in f^{-1}(p)} \text{Res}_q(\omega).$$

Proof. Let $v_X : \overline{X} \to X$ be the normalization of X. Let also $\overline{f} : \overline{X} \to Y$ be the natural lifting of f, that is $\overline{f} = f \circ v_X$.

Since $\overline{f}^{-1}(p) = (f \circ v_X)^{-1}(p)$ and because of definition (3.3), one verifies that the proposition holds for $f : X \to Y$ if it holds for $\overline{f} : \overline{X} \to Y$. It is therefore harmless to assume that both X and Y are smooth.

Let $q_1, \ldots, q_m$ be the pre-images of p under f. For $i \in \underline{m}$, let $f_i : (X, q_i) \to (Y, p)$ be the germ of analytic morphism induced by f, and let $\omega_i \in \mathcal{M}_{X,q_i}$ be the germ of ω at q_i.

Clearly, $\text{tr}_f(\omega) = \sum_{i=1}^m \text{tr}_{f_i}(\omega_i)$ as germs at p. By the additivity of the residue, it follows that

$$\text{Res}_p(\text{tr}_f(\omega)) = \sum_{i=1}^m \text{Res}_p(\text{tr}_{f_i}(\omega_i))$$

For each $i \in \underline{m}$, in a suitable local coordinate x_i, the germ of function f_i can be written $f_i(x_i) = x_i^{n_i}$ for a suitable positive integer n_i. A quick inspection of (3.1) leads to the identity $\text{Res}_p[\text{tr}_{f_i}(\omega_i)] = \text{Res}_{q_i}[\omega_i]$ for every $i \in \underline{m}$. The proposition follows. $\square$

3.2.2 Abelian Differentials

Let X be a reduced analytic curve and let $\nu : \overline{X} \to X$ be its desingularization. An
abelian differential ω on X is a global meromorphic 1-form on $\overline{X}$ which satisfies

$$\operatorname{Res}_p(f\omega) = \sum_{q \in \nu^{-1}(p)} \operatorname{Res}_q[\nu^*(f\omega)] = 0$$

for every $p \in X$ and every $f \in \mathcal{O}_{X,p}$.

For any open subset $U \subset X$, let $\omega_X(U)$ be the set of abelian differentials on U.
Of course $\omega_X(U)$ inherits from $\mathcal{M}_X(U)$ a structure of $\mathcal{O}_X(U)$-module. Actually,
even more is true: the subsheaf ω_X of $\mathcal{M}_X$ is coherent. A proof of this fact will be
presented later (cf. Sect. 3.2.4 below) under the assumption that X is contained in a
smooth surface. The general case will not be treated in this book. In the algebraic
category the result goes back to Rosenlicht [122]. For a treatment of the analytic
case see [10].

Remark 3.2.2. It is possible to characterize abelian differentials in terms of currents.
Indeed, a meromorphic 1-form ω on a curve X is abelian if and only if $\overline{\partial}[\omega] = 0$
where $[\omega]$ is a current associated with ω.[2] That is, if and only if

$$\langle \overline{\partial}[\omega], \theta \rangle := \langle [\omega], \overline{\partial}\theta \rangle = \int_X \omega \wedge \overline{\partial}\theta = 0,$$

for every smooth complex-valued function θ with compact support on X. The
characterization of abelian differentials in terms of currents generalizes promptly
and allows to define a notion of abelian differential k-forms in arbitrary dimension.
The interested reader can consult the references [10, 79].

For an arbitrary reduced projective curve X, the coherence of ω_X implies that
$H^0(X, \omega_X)$ is a finite dimensional vector space. Its dimension is, by definition, the
arithmetic genus $g_a(X)$ of X. The **geometric genus** $g(X)$ of X is in turn defined
as the dimension of $H^0(\overline{X}, \Omega^1_{\overline{X}})$, where $\overline{X}$ is the desingularization of X.

When X is smooth, the sheaf ω_X is nothing else than Ω^1_X, hence $g(X)$ coincides
with $g_a(X)$. For singular curves, an equality between $g(X)$ and $g_a(X)$ is the
exception rather than the rule.

The sheaf ω_X is also called the **dualizing sheaf** of X. The terminology comes
from Serre's duality for projective curves: *for any coherent sheaf $\mathscr{F}$ on a projective
curve X, there are natural isomorphisms between $H^i(X, \mathscr{F})$ and $H^{1-i}(X, \mathscr{F}^* \otimes \omega_X)^*$ for $i = 0, 1$.*

---

[2]There is a subtlety here: a priori, $[\omega]$ is not intrinsically defined. According to [10], it is only
defined up to the addition of a $\overline{\partial}$-closed current of type $(1, 0)$ supported on the (finite) singular set
of X (cf. [49] for a non-trivial example). In any case, the condition for $[\omega]$ to be $\overline{\partial}$-closed is well
defined and intrinsic.

When X is not just projective but also connected, a consequence of Serre's duality is **Riemann–Roch Theorem**: *for any line-bundle $\mathcal{L}$ on X, the identity $\chi(X, \mathcal{L}) = \deg(\mathcal{L}) - g_a(X) + 1$ holds true.*[3]

Applying Riemann–Roch Theorem to the dualizing sheaf itself, one obtains the **genus formula** for irreducible projective curves

$$\deg(\omega_X) = 2g_a(X) - 2. \tag{3.4}$$

Before proceeding toward the proof of Abel's Theorem for arbitrary curves, a couple of examples will be considered in order to clarify the notion of abelian differential.

Example 3.2.3. Let

$$X = \{(x, y) \in \mathbb{C}^2 \mid y^2 - x^3 = 0\}.$$

Clearly, it is a curve with the origin of $\mathbb{C}^2$ as its unique singular point. The stalk $\omega_{X,0}$ is a free $\mathcal{O}_{X,0}$-module generated by $y^{-1}dx$. Indeed, the normalization of X is given by

$$\nu : \mathbb{C} \longrightarrow X , \ t \longmapsto (t^2, t^3).$$

Therefore $\nu^* \mathcal{O}_{X,0} = \nu^* \mathcal{O}_{\mathbb{C}^2,0} = \nu^* \mathbb{C}\{x, y\} = \mathbb{C}\{t^2, t^3\}$. If ω is a meromorphic differential on X, then $\nu^* \omega = \sum_{i=-k}^{\infty} a_i t^i dt$ for a certain positive integer k. Moreover, if ω is abelian, then not only $a_{-1} = \mathrm{Res}_0(\nu^* \omega)$ must be zero but also $a_{2n+3m-1} = \mathrm{Res}_0(\nu^*(x^n y^m \omega))$ for any pair of positive integers (n, m). Hence every germ at 0 of abelian differential on X can be written as

$$(a_{-2} + a_0 t^2 + a_1 t^3 + \cdots) \frac{dt}{t^2} \in \mathbb{C}\{t^2, t^3\} \cdot \frac{dt}{t^2}.$$

In a similar vein, a family of rational projective curves contained in projective spaces of dimension $n \geq 2$ is considered below.

Example 3.2.4. Let us fix $n \geq 2$ and let C be the rational curve in $\mathbb{P}^n$ parametrized by

$$\nu : \mathbb{P}^1 \longrightarrow \mathbb{P}^n , \ [s : t] \longmapsto \left[s^{2n} : s^n t^n : s^{n-1} t^{n+1} : \cdots : st^{2n-1} : t^{2n}\right].$$

The curve C has degree $2n$, is singular, $p = [1 : 0 : \cdots : 0]$ is its unique singular point and the parametrization ν is its desingularization.

---

[3]In the formula, $\chi(X, \mathcal{L})$ stands for the Euler-characteristic of the line-bundle $\mathcal{L}$ which, by definition, is $h^0(X, \mathcal{L}) - h^1(X, \mathcal{L})$.

For any rational 1-form ω on C, its pull-back by ν on $\mathbb{P}^1$ writes $\nu^*\omega = f(t)dt$ in the coordinate $[1 : t]$, for a certain rational function $f(t) \in \mathbb{C}(t)$. Let us assume that ω is abelian. Since ω_C coincides with the sheaf of holomorphic differentials on $C_{sm} = C \setminus \{p\}$, the differential $\nu^*\omega = f(t)dt$ must be holomorphic on $\mathbb{C}^*$ as well as at infinity. Therefore $\nu^*\omega = t^{-a}dt$ for a certain integer $a > 1$.

It is an instructive exercise to show that ν^* induces an isomorphism

$$H^0(C, \omega_C) \simeq \left\langle \frac{dt}{t^2}, \frac{dt}{t^3}, \ldots, \frac{dt}{t^n}, \frac{dt}{t^{n+1}}, \frac{dt}{t^{2n+2}} \right\rangle.$$

Abelian Differentials and Traces

The following proposition can be seen as a first evidence that the concept of abelian differential is the appropriate one to extend Abel's addition Theorem to singular projective curves. Note that, as for Proposition 3.1.1, its converse does not hold true.

Proposition 3.2.5. *Let ω be an abelian differential on a reduced analytic curve X. If $f : X \to Y$ is a ramified covering onto a smooth curve Y, then $\mathrm{tr}_f(\omega)$ is a holomorphic differential on Y.*

Proof. The proof follows the same lines as that of Proposition 3.1.1 with some extra ingredients borrowed from the proof of Proposition 3.2.1. The reader is invited to fill in the details.   □

Beside the concept of abelian differential, the main extra ingredient used to generalize Abel's addition Theorem from smooth to arbitrary curves is the following characterization of abelian differentials in terms of their traces under linear projections.

Proposition 3.2.6. *Let $X \subset (\mathbb{C}^n, 0)$ be a germ of reduced analytic curve and ω be a meromorphic 1-form on X. The following assertions are equivalent:*

(a) ω is abelian;
(b) the trace $\mathrm{tr}_p(\omega)$ of ω relative to p is holomorphic for a generic linear projection
    $p : (\mathbb{C}^n, 0) \to (\mathbb{C}, 0)$.

Proof. From Proposition 3.2.5, it is clear that (a) implies (b). To prove that (b) implies (a), let us assume that ω is not holomorphic at 0. If this is the case, then there exists $f \in \mathcal{O}_{X,0}$ such that

$$\mathrm{Res}_0(f\omega) \neq 0.$$

By the additivity of the residue, the function f can be replaced by the restriction to X of a monomial function

$$x^J = x_1^{j_1} \cdots x_n^{j_n} \in \mathcal{O}_{\mathbb{C}^n,0} = \mathbb{C}\{x_1, \ldots, x_n\}$$

with $J = (j_1, \ldots, j_n) \in \mathbb{N}^n$, which still satisfies $\mathrm{Res}_0(x^J \omega) \neq 0$.

For $\epsilon = (\epsilon_1, \ldots, \epsilon_n) \in (\mathbb{C}, 1)^n$, let us define $p_\epsilon : (\mathbb{C}^n, 0) \to (\mathbb{C}, 0)$ as the linear projection

$$p_\epsilon(x_1, \ldots, x_n) = \sum_{i=1}^{n} \epsilon_i x_i .$$

Let us now consider the monomial $t^{|J|} \in \mathcal{O}_{\mathbb{C},0} = \mathbb{C}\{t\}$, where $|J| = \sum j_i$. It follows immediately from the algebraic definition of the trace (see Remark 3.1.2) that for every $\epsilon \in (\mathbb{C}, 1)^n$:

$$t^{|J|} \mathrm{tr}_{p_\epsilon}(\omega) = \mathrm{tr}_{p_\epsilon}\left(p_\epsilon^*(t^{|J|})\omega\right) .$$

Consequently, Proposition 3.2.1 implies

$$\mathrm{Res}_0\left(t^{|J|} \mathrm{tr}_{p_\epsilon}(\omega)\right) = \mathrm{Res}_0\left(p_\epsilon^*(t^{|J|})\omega\right) .$$

Using again the additivity of the residue, one gets

$$\mathrm{Res}_0\left(p_\epsilon^*(t^{|J|})\omega\right) = \sum_{\substack{K \in \mathbb{N}^n \\ |K| = |J|}} \binom{|K|}{K} \epsilon^K \mathrm{Res}_0(x^K \omega) , \tag{3.5}$$

where $\epsilon^K = \epsilon_1^{k_1} \cdots \epsilon_n^{k_n}$ and $\binom{|K|}{K} = \binom{|K|}{k_1} \cdots \binom{|K|}{k_n}$ for any $K = (k_1, \ldots, k_n) \in \mathbb{N}^n$.

Since $\mathrm{Res}_0(x^J \omega) \neq 0$, the polynomial in the variables $\epsilon_1, \ldots, \epsilon_m$ on the right-hand side of (3.5) is not zero. Consequently, for a generic ϵ, the meromorphic 1-form $\mathrm{tr}_{p_\epsilon}(p_\epsilon^*(t^{|J|})\omega)$ has a non-zero residue at the origin. This suffices to establish that (b) implies (a). □

3.2.3 Abel's Addition Theorem

Having the concept of abelian differential as well as Proposition 3.2.6 at hand, there is no difficulty to adapt the proof of Theorem 3.1.3 in order to establish Abel's addition Theorem for arbitrary algebraic projective curves.

Theorem 3.2.7 (Abel's Addition Theorem). *If ω is a meromorphic 1-form on a projective curve $C \subset \mathbb{P}^n$, then $\mathrm{Tr}(\omega)$ is a meromorphic 1-form on $\check{\mathbb{P}}^n$. Moreover, ω is abelian if and only if its trace $\mathrm{Tr}(\omega)$ vanishes identically.*

As a by-product, Theorem 3.1.5 also generalizes to the following:

Theorem 3.2.8. *If C is a projective curve of degree k and H_0 is a hyperplane intersecting it transversely, then the space of abelian 1-forms on C injects into the space of abelian relations of the dual web $\mathcal{W}_C(H_0)$.*

Using a notation similar to the one of Sect. 3.1.4, the preceding result can be formulated as follows: the injective linear map

$$H^0(C, \omega_C) \longrightarrow \left(\Omega^1(\check{\mathbb{P}}^n, H_0)\right)^k$$

$$\omega \longmapsto (p_1^* \omega, \dots, p_k^* \omega)$$

factors through $\mathcal{A}(\mathcal{W}_C(H_0))$.

Of course there is also a corresponding version of Castelnuovo's bound for arbitrary reduced curves which intersect a generic hyperplane in points in general position.

Theorem 3.2.9. *Let C be a reduced projective curve on $\mathbb{P}^n$ of degree k. If the dual web is generically smooth, for instance if C is irreducible and non-degenerate, then $h^0(C, \omega_C) \leq \pi(n, k)$.*

To ease further reference to projective curves with generically smooth dual web, these will be labeled $\mathcal{W}$-**generic curves**. Notice that according to Proposition 1.4.5, an irreducible curve is $\mathcal{W}$-generic if and only if it is non-degenerate.

3.2.4 Abelian Differentials for Curves on Surfaces

Let now X be a reduced curve on a smooth compact connected surface S. The purpose of this section is to describe the sheaf ω_X in terms of sheaves over S. As usual, K_S denotes the sheaf of holomorphic 2-forms on S—the **canonical sheaf** of S—and $K_S(X)$ is used as an abbreviation of $K_S \otimes \mathcal{O}_S(X)$.

If U is a sufficiently small neighborhood of a point $p \in X$, then $X \cap U = \{f = 0\}$ for some $f \in \mathcal{O}_S(U)$ generating the ideal $\mathcal{I}_X(U)$. Notice that any section $\eta \in \Gamma(U, K_S(X))$ can be written as

$$\eta = f^{-1} h \, dx \wedge dy$$

with $h \in \mathcal{O}_S(U)$, for some coordinate functions x and y on U.

If the latter are not constant on any of the irreducible components of X, then $\mathrm{Res}_X(\eta)$—the **residue of η along X**—is, by definition, the restriction to X of the meromorphic 1-form $(h/\partial_y f)dx$. Explicitly,

$$\mathrm{Res}_X(\eta) = \left(\frac{h \, dx}{\partial f / \partial y}\right)\Big|_X.$$

It is easy to verify that $\mathrm{Res}_X(\cdot)$ does not depend on the choice of f nor on the choice of the local coordinates x, y. In the literature, $\mathrm{Res}_X(\eta)$ also appears under the label of Poincaré's, as well as Leray's, residue of η along X.

Notice that for every function $g \in \mathcal{O}_S(U)$ and every 2-form η as above, one has

$$\mathrm{Res}_X(g \cdot \eta) = (g|_X) \cdot \mathrm{Res}_X(\eta).$$

Thus the map Res_X can be interpreted as a morphism of $\mathcal{O}_S$-modules from $K_S(X)$ to $\mathcal{M}_X$, the sheaf of meromorphic differential 1-forms on X. Of course, the structure of $\mathcal{O}_S$-module on the latter sheaf is the one induced by the inclusion of X into S.

Clearly, the kernel of $\mathrm{Res}_X : K_S(X) \to \mathcal{M}_X$ coincides with the natural inclusion of the canonical sheaf K_S into $K_S(X)$. Therefore the sequence of sheaves

$$0 \longrightarrow K_S \longrightarrow K_S(X) \xrightarrow{\ \mathrm{Res}_X\ } \mathrm{Im}\,\mathrm{Res}_X \subset \mathcal{M}_X$$

is exact.

Proposition 3.2.10. *The image of* Res_X *is exactly* ω_X, *the sheaf of abelian differentials on* X. *Consequently* ω_X *is* $\mathcal{O}_X$-*coherent and the adjunction formula*

$$\left(K_S \otimes \mathcal{O}_S(X)\right)\big|_X \simeq \omega_X$$

holds true.

The proof of Proposition 3.2.10 will make use of the following lemma.

Lemma 3.2.11. *Let* $X = \{f = 0\}$ *be a reduced analytic curve defined at a neighborhood of the origin of* $\mathbb{C}^2$. *Let us suppose that the coordinate functions* x, y *are not constant on any of the irreducible components of* X. *If* $\mathbb{S}$ *is a sufficiently small sphere centered at the origin, and transverse to* X, *then the identity*

$$\frac{1}{2\pi i} \int_{\mathbb{S} \cap X} \frac{h\,dx}{\partial_y f} = \lim_{\epsilon \to 0} \int_{\mathbb{S} \cap \{|f| = \epsilon\}} \frac{h\,dx \wedge dy}{f}$$

is valid for any meromorphic function h *on* $(\mathbb{C}^2, 0)$.

We refer to [11, p. 52] for a proof.

Proof of Proposition 3.2.10. Clearly $\mathrm{Im}(\mathrm{Res}_X) = \omega_X$ is a local statement. Moreover, for a smooth point p of X, it is not difficult to see that both $\omega_{X,p}$ and $\mathrm{Im}(\mathrm{Res}_X)_p$ are isomorphic to $\Omega^1_{X,p}$. Let $p \in X$ be a singular point and let (x, y) be local coordinates centered at p satisfying the assumption of Lemma 3.2.11.

To prove that $\mathrm{Im}(\mathrm{Res}_X)_p$ is contained in $\omega_{X,p}$ notice that the former $\mathcal{O}_{S,p}$-module is generated by $(\partial_y f)^{-1}dx = \mathrm{Res}_X(f^{-1}dx \wedge dy)$. It follows first from (3.2) and then from Lemma 3.2.11 that for every $h \in \mathcal{O}_{S,p}$, one has

$$\mathrm{Res}_0\left(\frac{h\,dx}{\partial_y f}\right) = \frac{1}{2\pi i}\int_{S\cap X}\frac{h\,dx}{\partial_y f} = \lim_{\epsilon\to 0}\int_{S\cap\{|f|=\epsilon\}}\frac{h\,dx\wedge dy}{f}.$$

But, by Stoke's Theorem, for every sufficiently small $\epsilon > 0$:

$$\int_{S\cap\{|f|=\epsilon\}}\frac{h\,dx\wedge dy}{f} = \int_{S\cap\{|f|\geq\epsilon\}}d\left(\frac{h\,dx\wedge dy}{f}\right) = 0.$$

Therefore $(\partial_y f)^{-1}dx \in \omega_{X,p}$ as wanted.

Suppose now that the 1-form $\eta = h(\partial_y f)^{-1}dx$ is abelian for some meromorphic function h on X. Let $n \in \mathbb{N}$ be the smallest integer for which the function $x^n h$ is holomorphic at p, that is belongs to $\mathcal{O}_{X,p}$. Let $h_n \in \mathcal{O}_{S,p}$ be a holomorphic function whose restriction to X is equal to $x^n h$.

If $n = 0$, then the relation $\eta = \mathrm{Res}_X(h_0 f^{-1}dx\wedge dy)$ with $h_0 \in \mathcal{O}_{S,p}$ shows that $\eta \in \mathrm{Im}(\mathrm{Res}_X)_p$ as requested. Thus, from now on, n will be assumed positive.

Since η is abelian, one has

$$0 = \mathrm{Res}_0(gx^{n-1}\eta) = \mathrm{Res}_0\left(g\frac{h_n}{x}\frac{dx}{\partial_y f}\right)$$

for every $g \in \mathcal{O}_{S,p}$. Applying Lemma 3.2.11, and then Stoke's Theorem, one deduces that the right-hand side is, up to multiplication by $2\pi i$, equal to

$$\lim_{\epsilon\to 0}\int_{S\cap\{|f|=\epsilon\}}\left(g\frac{h_n}{x}\frac{dx\wedge dy}{f}\right) = -\lim_{\epsilon\to 0}\int_{S\cap\{|x|=\epsilon\}}\left(g\frac{h_n}{x}\frac{dx\wedge dy}{f}\right).$$

Applying Lemma 3.2.11 again, but now to the curve $Y = \{x = 0\}$, yields

$$\mathrm{Res}_0\left(\frac{gh_n}{f}dy\right) = 0$$

for every $g \in \mathcal{O}_{S,p}$. But this implies that h_n/f is a holomorphic function on Y. Therefore

$$h_n = h_{n-1}x + af$$

where $a, h_{n-1} \in \mathcal{O}_{S,p}$ are holomorphic functions. Thus on X

$$x^{n-1}h = h_{n-1},$$

which contradicts the minimality of n. The inclusion $\omega_{X,p} \subset \mathrm{Im}(\mathrm{Res}_X)_p$ follows. Therefore $\omega_X = \mathrm{Im}(\mathrm{Res}_X)$. The coherence of the sheaf ω_X and adjunction formula follow at once from the exact sequence

$$0 \to K_S \to K_S(X) \to \omega_X \to 0.$$

$\square$

The preceding proof also shows the following:

Corollary 3.2.12. *For a curve X embedded in a smooth surface the sheaf ω_X is locally free.*

Curves for which ω_X is locally free are usually called **Gorenstein curves**. Corollary 3.2.12 can be succinctly rephrased as: *germs of planar curves are Gorenstein.* This is no longer true for arbitrary singularities. The simplest example is perhaps the germ of curve X on $(\mathbb{C}^3, 0)$ having the three coordinate axis as irreducible components.

Corollary 3.2.12 and the genus formula (3.4) imply the following:

Corollary 3.2.13. *If X is an irreducible projective curve embedded in a smooth compact surface S, then*

$$g_a(X) = \frac{(K_S + X) \cdot X}{2} + 1 . \tag{3.6}$$

Abelian Differentials on Planar Curves

The results just presented for arbitrary smooth surfaces S will be now specialized to the projective plane $\mathbb{P}^2$.

Let $C \subset \mathbb{P}^2$ be a reduced algebraic curve of degree k and let $\mathcal{I}_C$ be its ideal sheaf. Since $K_{\mathbb{P}^2} = \mathcal{O}_{\mathbb{P}^2}(-3)$ and $\mathcal{O}_{\mathbb{P}^2}(C) = \mathcal{O}_{\mathbb{P}^2}(k)$, the adjunction formula reads as

$$\omega_C = \mathcal{O}_{\mathbb{P}^2}(k - 3)|_C .$$

Because $\mathcal{I}_C(k - 3)$ is isomorphic to $\mathcal{O}_{\mathbb{P}^2}(-3)$, both cohomology groups $H^0(\mathbb{P}^2, \mathcal{I}_C(k-3))$ and $H^1(\mathbb{P}^2, \mathcal{I}_C(k-3))$ are trivial. Therefore, the restriction map

$$H^0\big(\mathbb{P}^2, \mathcal{O}_{\mathbb{P}^2}(k - 3)\big) \longrightarrow H^0\big(C, \mathcal{O}_C(k - 3)\big)$$

is an isomorphism. Combined with the adjunction formula, this yields

$$H^0\big(\mathbb{P}^2, K_{\mathbb{P}^2}(C)\big) \simeq H^0\big(\mathbb{P}^2, \mathcal{O}_{\mathbb{P}^2}(k - 3)\big) \simeq H^0(C, \omega_C).$$

The isomorphism from the leftmost to the rightmost group is induced by Res_C.

The discussion above is summarized, and made more explicit, in the following:

Corollary 3.2.14. *Let $\{f(x, y) = 0\}$ be a reduced equation for C in generic affine coordinates x, y on $\mathbb{P}^2$. Then*

$$H^0(C, \omega_C) \simeq \left\langle \frac{p(x, y)dx}{\partial f / \partial y} \;\middle|\; \begin{array}{c} p \in \mathbb{C}[x, y] \\ \deg p \leq k - 3 \end{array} \right\rangle .$$

Consequently $g_a(C) = \frac{(k-1)(k-2)}{2}$.

3.3   Algebraic Webs of Maximal Rank

In view of Theorem 3.2.8, it suffices to consider $\mathcal{W}$-generic curves $C \subset \mathbb{P}^n$ with $h^0(C, \omega_C)$ attaining Castelnuovo's bound $\pi(n, \deg(C))$ in order to have examples of webs of maximal rank. Classically, a degree k irreducible non-degenerate curve $C \subset \mathbb{P}^n$ such that $g_a(C) = \pi(n, k) > 0$ is called a **Castelnuovo curve**. Notice that the definition implies that a Castelnuovo curve has necessarily degree $k \geq n + 1$, since otherwise $g_a(C) = 0$ according to Theorem 3.1.7.

The simplest examples of Castelnuovo curves are the irreducible planar curves of degree at least three. Since the arithmetic genus of a reduced planar curve C is $\pi(2, \deg(C))$, such curves are certainly Castelnuovo curves.

In sharp contrast, when the ambient dimension is at least three, the Castelnuovo curves are the exception rather than the rule. Indeed, the analysis carried out in Sect. 2.4 to control the geometry of the conormals of maximal rank webs was originally developed by Castelnuovo to control the geometry of hyperplane sections of Castelnuovo curves. Reformulating Proposition 2.4.10 in terms of curves, instead of webs, provides the following:

Proposition 3.3.1. *If $C \subset \mathbb{P}^n$ is a Castelnuovo curve of degree $k \geq 2n + 1$, then a generic hyperplane section of C is contained in a rational normal curve of degree $n - 1$.*

Actually Castelnuovo's Theory goes further and says that a Castelnuovo curve $C \subset \mathbb{P}^n$ is contained in a surface S, which is cut out by the linear system $|I_C(2)|$ of hyperquadrics containing C, which has rational normal curves as generic hyperplane sections. Because the degree of S is the same as the one of a generic hyperplane section and rational normal curves in $\mathbb{P}^{n-1}$ have degree $n - 1$, the surface S is of minimal degree. Theorem 2.4.13 applies and implies that S is either a plane, a Veronese surface in $\mathbb{P}^5$, or a rational normal scroll $S_{a,b}$ with $a + b = n - 1$.

The proof of these results will not be presented here,[4] but the rather easier determination of Castelnuovo curves in surfaces of the above list will be sketched below.

3.3.1   Castelnuovo Curves on the Veronese Surface

Let $S \subset \mathbb{P}^5$ be the Veronese surface. Recall that S is the embedding of $\mathbb{P}^2$ into $\mathbb{P}^5$ induced by the complete linear system of conics $|\mathcal{O}_{\mathbb{P}^2}(2)|$. If C is an irreducible curve contained in S, then its degree k as a curve in $\mathbb{P}^5$ is twice its degree d as a curve in $\mathbb{P}^2$.

On the one hand, one has

---

[4]The interested reader may consult, for instance, [7, 67], or [68].

$$h^0(C, \omega_C) = \frac{(d-1)(d-2)}{2} = \frac{(k-2)(k-4)}{8}.$$

On the other hand, Castelnuovo's bound predicts that

$$h^0(C, \omega_C) \leq \begin{cases} (e-1)(2e-1) & \text{if } k = 4e \\ e(2e-1) & \text{if } k = 4e+2. \end{cases}$$

Therefore, both cases lead to Castelnuovo curves.

3.3.2 Castelnuovo Curves on Rational Normal Scrolls

Let S be a rational normal scroll $S_{a,b}$ with $0 \leq a \leq b$ and $a + b = n - 1$. If S is singular, that is if $a = 0$, let us replace S by its desingularization $\mathbb{P}(\mathcal{O}_{\mathbb{P}^1} \oplus \mathcal{O}_{\mathbb{P}^1}(b))$ which will still be denoted by S.

The Picard group of S is the free $\mathbb{Z}$-module of rank 2 generated by H, the class of a hyperplane section, and L, the class of a line of the ruling. Of course,

$$H^2 = n - 1, \quad H \cdot L = 1 \quad \text{and} \quad L^2 = 0.$$

If one writes $K_S = \alpha H + \beta L$, then the two integers α and β can be determined using the genus formula (3.6). Since both H and L are smooth rational curves, one has

$$-2 = 2g(H) - 2 = H^2 + K_S \cdot H = (\alpha + 1)(n - 1) + \beta,$$

$$\text{and} \quad -2 = 2g(L) - 2 = L^2 + K_S \cdot L = \alpha.$$

Therefore $\alpha = -2$ and $\beta = n - 3$, that is

$$K_S = -2H + (n - 3)L.$$

Let α and β be new integer constants unrelated to the ones above. Let also C be an irreducible curve contained in S, numerically equivalent to $\alpha H + \beta L$. Notice that $\deg(C) = C \cdot H = \alpha(n - 1) + \beta$. The genus formula (3.6) implies

$$g_a(C) = \frac{C^2 + K_S \cdot C}{2} + 1 = \frac{1}{2}(\alpha - 1)\Big((n - 1)\alpha + 2(\beta - 1)\Big).$$

Suppose now that the degree of C is equal to $k \geq n + 1$ and write

$$k - 1 = m(n - 1) + \epsilon,$$

as in Remark 2.3.9, that is with $m = \lfloor \frac{k-1}{n-1} \rfloor$ and $\epsilon \in \{0, \ldots, n - 2\}$.

If C is Castelnuvo, then

$$g_a(C) = \binom{m}{2}(n-1) + m\epsilon.$$

Thus the Castelnuovo's curves on S provide solutions to the following system of equations

$$\deg(C) = m(n-1) + \epsilon + 1 = \alpha(n-1) + \beta,$$

$$g_a(C) = \binom{m}{2}(n-1) + m\epsilon = \frac{1}{2}(\alpha-1)\big((n-1)\alpha + 2(\beta-1)\big),$$

subject to the arithmetical constraints $m, n \geq 0$, and $0 \leq \epsilon \leq n-2$; and the geometrical constraint $\alpha = C \cdot L > 0$.

It can be shown that the only possible solutions are

$$(\alpha, \beta) = \big(m+1, -(n-1-\epsilon)\big)$$

or $(\alpha, \beta) = (m, 1)$, the latter case only when $\epsilon = 0$.

If $\mathcal{L} = \mathcal{O}_S(\alpha H + \beta L)$, with (α, β) being one of the solutions above, then Riemann–Roch's Theorem for surfaces, see, for instance, [70, Theorem 1.6, Chapter V], implies

$$\chi(\mathcal{L}) = \frac{1}{2}(\alpha H + \beta L)\big((\alpha - 2)H + (\beta + n - 3)L\big) + \chi(S)$$

$$= \binom{\alpha + 1}{2}(n-1) + (\alpha + 1)(\beta + 1).$$

Notice that

$$K_S \otimes \mathcal{L}^* = \mathcal{O}_S\big((-2 - \alpha)H + (n - 3 - \beta)L\big).$$

Because $\alpha > 0$, the following inequality holds true

$$(K_S \otimes \mathcal{L}^*) \cdot L < 0.$$

Thus $h^0(S, K_S \otimes \mathcal{L}^*) = 0$. Serre's duality for surfaces implies the same for $h^2(S, \mathcal{L})$, that is $h^2(S, \mathcal{L}) = 0$. Consequently $h^0(S, \mathcal{L}) \geq \chi(\mathcal{L})$. Moreover, if $\alpha > 2$, then $h^0(\mathcal{L}) \geq \chi(\mathcal{L}) > 0$.

Since α is $\lfloor \frac{k-1}{n-1} \rfloor$ or $\lfloor \frac{k-1}{n-1} \rfloor + 1$, the linear system $|\mathcal{L}|$ is non-empty when $k \geq 2n$.

One can push this analysis further and show that the linear system $|\mathcal{L}|$ is base-point free whenever $k \geq 2n$. As a consequence, the generic element of $|\mathcal{L}|$ is an irreducible smooth curve according to Bertini's Theorem. See [67] for details.

The discussion above is summarized in the following statement.

Proposition 3.3.2. *For any integer $n \geq 3$, any pair (a, b) of non-negative integers summing up to $n - 1$, and any $k \geq 2n$, there exist Castelnuovo curves of degree k contained in $S_{a,b} \subset \mathbb{P}^n$.*

3.3.3 Webs of Maximal Rank

From what has been said in the previous sections, the following result follows promptly.

Proposition 3.3.3. *For any integers $n \geq 2$ and $k \geq 2n$, there exist smooth k-webs of maximal rank on $(\mathbb{C}^n, 0)$. These are the algebraic webs of the form $\mathcal{W}_C(H_0)$, where C is a degree k Castelnuovo curve in $\mathbb{P}^n$ and H_0 a hyperplane intersecting it transversely.*

It remains to discuss smooth k-webs of maximal rank on $(\mathbb{C}^n, 0)$ with $k < 2n$.

For $k \leq n$ there is not much to say since smooth k-webs on $(\mathbb{C}^n, 0)$ have always rank 0 and are equivalent to $\mathcal{W}(x_1, \ldots, x_k)$.

For $k = n + 1$, a web of maximal rank carries exactly one non-zero abelian relation because $\pi(n, n + 1) = 1$. Thus, if

$$u_1, \ldots, u_{n+1} : (\mathbb{C}^n, 0) \to \mathbb{C}$$

are submersions defining $\mathcal{W}$, then its unique abelian relation takes the form $f_1(u_1) + \cdots + f_{n+1}(u_{n+1}) = 0$ for suitable holomorphic germs $f_i : (\mathbb{C}, 0) \to (\mathbb{C}, 0)$. It follows that $\mathcal{W}$ is equivalent to the parallel $(n+1)$-web $\mathcal{W}(x_1, \ldots, x_n, x_1 + \cdots + x_n)$.

For $k \in \{n + 2, \ldots, 2n - 1\}$, it is fairly simple to construct smooth k-webs of maximal rank $\pi(n, k) = k - n$. It suffices to consider submersions $u_1, \ldots, u_k : (\mathbb{C}^n, 0) \to (\mathbb{C}, 0)$ of the form:

$$u_i(x_1, \ldots, x_n) = x_i \qquad \text{for } i \in \underline{n}$$

$$\text{and} \quad u_i(x_1, \ldots, x_n) = \sum_{j=1}^{n} u_i^{(j)}(x_j) \qquad \text{for } i \in \{n + 1, \ldots, k\}.$$

Here $u_i^{(j)} : (\mathbb{C}, 0) \to (\mathbb{C}, 0)$ are germs of submersions in one variable. It is clear that for a generic choice of the submersions $u_i^{(j)}$, the k-web $\mathcal{W} = \mathcal{W}(u_1, \ldots, u_k)$ is smooth. Moreover, the definition of u_i when $i > n$ can be interpreted as an abelian relation of $\mathcal{W}$. Therefore $\text{rank}(\mathcal{W}) \geq k - n$. But $\pi(n, k) = k - n$, therefore $\mathcal{W}$ is of maximal rank.

§

The examples of k-webs of maximal rank with $k = n + 1$ or $k \geq 2n$ are of a different nature than the ones presented for $k \in \{n + 2, \dots, 2n - 1\}$. While the equivalence classes of the former examples belong to finite dimensional families, the ones of the latter belong to infinite dimensional families, as can be easily verified. Although this could be just a coincidence, the main result of this book is that this is not case at least when $n \geq 3$ and $k \geq 2n$. A precise statement will be given in Chap. 5.

Remark 3.3.4. If $C \subset \mathbb{P}^n$ is a non-degenerate, but not $\mathcal{W}$-generic, curve of degree k, then it may happen that $h^0(\omega_C) > \pi(n, k)$. According to Proposition 1.4.6, such a curve C cannot be irreducible. There are works (see [9, 71, 127]) showing the existence of a function $\tilde{\pi}(n, k)$ which bounds the arithmetical genus of non-degenerate projective curves in $\mathbb{P}^n$ of degree k. Of course, $\tilde{\pi}(n, k)$ is greater than Castelnuovo's number $\pi(n, k)$. Moreover, the curves C attaining this bound have been classified. They all have a plane curve among their irreducible components.

3.4   Webs and Families of Hypersurfaces

The construction of the dual web $\mathcal{W}_C$ of a projective curve C, as well as the definition of the trace relative to the family of hyperplanes, makes use of the incidence variety $\mathcal{I} \subset \mathbb{P}^n \times \check{\mathbb{P}}^n$. The interpretation of $\mathcal{I}$ as the family of hyperplanes in $\mathbb{P}^n$ suggests the extension of both constructions to other families of hypersurfaces.

In this section, one such extension is described, and used in combination with Chern's bound for the rank of smooth webs to obtain bounds for the geometric genus of curves on abelian varieties.

3.4.1   Dual Webs with Respect to a Family

Let X and T be projective manifolds, and $\pi_T : X \times T \to T$, $\pi_X : X \times T \to X$ be the natural projections. Consider $\mathscr{X}$, a **family of hypersurfaces** in X parametrized by T. By definition $\mathscr{X}$ is an irreducible subvariety of $X \times T$ for which $\mathscr{X}_H = \pi_T^{-1}(H) \cap \mathscr{X}$ is a hypersurface of $X \times \{H\}$ for every $H \in T$ generic enough. It will be convenient, as it has been done with the family of hyperplanes on $\mathbb{P}^n$, to think of H as a point of T ($H \in T$), as well as a hypersurface in X ($H \subset X$).

Let $C \subset X$ be a reduced curve and $H_0 \subset X$ be a hypersurface which belongs to the family $\mathscr{X}$—that is, $H_0 \subset X$ is equal to $\pi_X(\mathscr{X}_{H_0})$ with $H_0 \in T$ being the corresponding point.

If $H_0 \subset X$ intersects C transversely in k distinct points then, as for the family of hyperplanes in $\mathbb{P}^n$, there are germs of holomorphic maps

$$p_i : (T, H_0) \to C$$

for $i \in \underline{k}$, implicitly defined by

$$C \cap \pi_X(\mathscr{X}_H) = \sum_{i=1}^{k} p_i(H).$$

It may happen that one of the points $p_i(H_0)$ is common to all the hypersurfaces in the family. It also may happen that for some pair $i, j \in \underline{k}$, the functions p_i and p_j define the same foliation. But, if the maps p_i are non-constant and define pairwise distinct foliations, then there is a naturally defined germ of (possibly singular) k-web on (T, H_0): $\mathcal{W}(p_1, \ldots, p_k)$. It will be called the $\mathscr{X}$-**dual web of** C **at** $H_0 \in T$, and denoted by $\mathcal{W}_C^{\mathscr{X}}(H_0)$.

If it is possible to define the germ of k-web $\mathcal{W}_C^{\mathscr{X}}(H_0)$ at any generic $H_0 \in T$, then there is no obstruction to define the global k-web $\mathcal{W}_C^{\mathscr{X}}$, the $\mathscr{X}$-**dual web of** C.

3.4.2 Trace Relative to Families of Hypersurfaces

For a curve $C \subset X$ and a meromorphic 1-form ω on C, it is possible to succinctly define the $\mathscr{X}$-**trace of** ω **relative to the family** $\mathscr{X}$ by the formula

$$\mathrm{Tr}_{\mathscr{X}}(\omega) = (\pi_T)_* (\pi_X)^*(\omega).$$

To give a sense to this expression, it is necessary to assume that a generic hypersurface in the family $\mathscr{X}$ intersects C in at most a finite number of points. In other words, no irreducible component of C is contained in the generic hypersurface of $\mathscr{X}$. If this is the case, consider then Σ_0, one of the irreducible components of $\pi_X^{-1}(C) \cap \mathscr{X}$ with dominant projection to T. Because a generic $H \in T$ intersects C in finitely many points, Σ_0 has the same dimension as T.

Let $\Sigma \to \Sigma_0$ be a resolution of singularities, and let us still denote by π_X, π_T the compositions of the natural projections with the resolution morphism. Then $\pi_X^* \omega$ is a meromorphic 1-form on Σ, and for a generic $H \in T$ there are local inverses for $\pi_T : \Sigma \to T$. Proceeding exactly as for the family of hyperplanes in $\mathbb{P}^n$, one can define a meromorphic 1-form η_{Σ_0} on the complement of the critical values of $\pi_T : \Sigma \to T$.

To extend η_{Σ_0} through the critical value set—or discriminant—of π_T, observe that outside a codimension two subset, the discriminant is smooth; the fibers over points in it are finite; and locally, at each pre-image of such a point, π_T is conjugated to a map of the form $(x_1, \ldots, x_n) \mapsto (x_1^r, x_2, \ldots, x_n)$ for some suitable positive integer r. Hence, as for the family of hyperplanes, η_{Σ_0} can be extended meromorphically through this set. But a meromorphic 1-form defined on the complement of a codimension two subset extends to the whole ambient space according to Hartog's extension theorem.

The $\mathscr{X}$-trace $\mathrm{Tr}_{\mathscr{X}}(\omega)$ is then defined as the sum of the meromorphic 1-forms η_{Σ_0} for Σ_0 ranging over all the irreducible components of $\pi_X^{-1}(C) \cap \mathscr{X}$ dominating T.

It is not hard to prove the following weak analogue of Abel's addition Theorem.

Proposition 3.4.1. *Let $\mathscr{X}$ be a family of hypersurfaces of X over a smooth projective variety T. If $C \subset X$ is a curve intersecting a generic hypersurface of the family at finitely many points, and if ω is a meromorphic 1-form on C, then the $\mathscr{X}$-trace of ω is a meromorphic 1-form on T. Moreover, if the pull-back of ω to a desingularization of C is holomorphic, then its $\mathscr{X}$-trace is a holomorphic 1-form on T.*

If $\mathscr{X}$ is the family of all hypersurfaces of degree d of $\mathbb{P}^n$, then a strong analogue of Abel's Theorem is valid: the $\mathscr{X}$-trace of ω is holomorphic if and only if ω is abelian. For arbitrary families of hypersurfaces, the equivalence between holomorphic $\mathscr{X}$-traces and abelian differentials is too much to be hoped for.

3.4.3 Bounds for Rank and Genus

From Proposition 3.4.1, one can obtain lower bounds for the rank of $\mathscr{X}$-dual webs for curves in X.

Proposition 3.4.2. *Let $C \subset X$ be a curve with $\mathscr{X}$-degree k. If $\mathcal{W}_C^{\mathscr{X}}$ is a k-web, then*

$$\mathrm{rank}\left(\mathcal{W}_C^{\mathscr{X}}\right) \geq g(C) - h^0\left(T, \Omega_T^1\right).$$

Proof. If $\overline{C}$ is a desingularization of C then, according to Proposition 3.4.1, the $\mathscr{X}$-trace of any holomorphic 1-form in $H^0(\overline{C}, \Omega_{\overline{C}}^1)$ is a holomorphic 1-form on T. Moreover, the map $\omega \longmapsto \mathrm{Tr}_{\mathscr{X}}(\omega)$ is a linear map from $H^0(\overline{C}, \Omega_{\overline{C}}^1)$ to $H^0(T, \Omega_T^1)$ whose kernel can be identified with a linear subspace of $\mathcal{A}(\mathcal{W}_C^{\mathscr{X}}(H_0))$ for a generic $H_0 \in T$. Since, by definition $g(C) = h^0(\Omega_{\overline{C}}^1)$, the proposition follows. $\square$

In analogy with the standard case of families of hyperplanes, Proposition 3.4.2 read backwards provides a bound for the genus of curves $C \subset X$ as soon as the k-web $\mathcal{W}_C^{\mathscr{X}}$ is generically smooth.

Proposition 3.4.3. *Let $C \subset X$ be a curve with $\mathscr{X}$-degree k. If $\mathcal{W}_C^{\mathscr{X}}$ is a generically smooth k-web, then*

$$g(C) \leq \pi(\dim T, k) + h^0\left(T, \Omega_T^1\right).$$

3.4.4 Families of Theta Translates

Let A be an abelian variety of dimension n and let $H_0 \subset A$ be an irreducible divisor in A. Recall that A acts on itself by translations and assume that H_0 has a finite isotropy group under this action.

Consider a second copy of A and denote it by $\check{A}$. Let $\mathscr{X}$ be the family of translates of $H_0 \subset A$ by points of $\check{A}$, that is

$$\mathscr{X} = \left\{ (x, y) \in A \times \check{A} \,\middle|\, x - y \in H_0 \right\}.$$

The natural projections from $\mathscr{X}$ to A and $\check{A}$ will be denoted by π and $\check{\pi}$, respectively.

Lemma 3.4.4. *Let A, H_0, and $\mathscr{X}$ be as above. If $C \subset A$ is an irreducible curve not contained in a translate of H_0, then $\mathcal{W}_C^{\mathscr{X}}$ is generically smooth.*

Proof. Let H_0 be a generic hypersurface in the family, intersecting C transversely in k points. At a neighborhood of H_0, write $C = C_1 \cup \cdots \cup C_k$. For simplicity, assume that $k = H_0 \cdot C \geq n$.

To prove that $\mathcal{W}_C^{\mathscr{X}}$ is generically smooth, first notice that

$$P = (p_1, p_2, \ldots, p_k) : (A, H_0) \longrightarrow C^k$$
$$H \longmapsto (H \cap C_1, \ldots, H \cap C_k)$$

has finite fibers. If not the fiber over $P(H_0)$ has positive dimension. Consequently one of the irreducible components of its Zariski closure is an analytic subset $Z \subset \check{A}$ of positive dimension. Clearly for every $H \in Z \subset \check{A}$ the intersection $H \cap C \subset A$ contains $H_0 \cap C$. Since H_0 has finite isotropy group $\pi(\check{\pi}^{-1}(Z))$ is equal to A. Therefore given an arbitrary point $q \in C$, there exists $H_q \in Z$ containing q. If C is not contained in H_q, then $H_q \cdot C > k$. But H_q and H_0 are algebraically equivalent, thus $k > H_q \cdot C = H_0 \cdot C = k$. This contradiction shows that P has finite fibers.

Using the irreducibility of C, one can show that the map $P_I = (p_{i_1}, \ldots, p_{i_n})$: $(A, H_0) \rightarrow C^n$ has finite fibers for any subset $I = \{i_1, \ldots, i_n\}$ of $\underline{k}$ having cardinality n. The reader is invited to fill in the details.

Since P_I has finite fibers and $\mathcal{W}(p_{i_1}, \ldots, p_{i_n})$ is an arbitrary n-subweb of $\mathcal{W}_C^{\mathscr{X}}$ at H_0, it follows that the web $\mathcal{W}_C^{\mathscr{X}}$ is generically smooth. $\qquad\square$

This lemma together with Proposition 3.4.3 promptly implies the following

Theorem 3.4.5. *Let $C \subset A$ be a curve with $\mathscr{X}$-degree k. If C is irreducible and is not contained in any translate of H_0, then*

$$g(C) \leq \pi(n, k) + n. \tag{3.7}$$

The most natural example of a pair (A, H_0) satisfying the above conditions is an irreducible principally polarized abelian variety (A, Θ). There is a version of Castelnuovo's Theory in this context, developed by Pareschi and Popa in [102]. In particular, they obtain the following bound for the genus of curves in A.

Theorem 3.4.6. *Let (A, Θ) be an irreducible principally polarized abelian variety of dimension n. Let $C \subset A$ be a non-degenerate[5] irreducible curve of degree $k = C \cdot \Theta$ in A. Let $m = \lfloor \frac{k-1}{n} \rfloor$, so that $k - 1 = mn + \epsilon$, with $0 \le \epsilon < n$. Then*

$$g(C) \le \binom{m+1}{2} n + (m+1)\epsilon + 1. \tag{3.8}$$

Moreover, the inequality is strict for $n \ge 3$ and $k \ge n + 2$.

For a fixed n, the bounds (3.7) and (3.8) are asymptotically equal to $\frac{k^2}{2(n-1)}$ and $\frac{k^2}{2n}$, respectively. The bound provided by Theorem 3.4.5 is asymptotically worse than the one provided by Theorem 3.4.6. But when $n \ge 3$ and for comparably small values of k, the former is sharper than the latter. Indeed, it can be verified that the bound of Theorem 3.4.5 is sharper than the one of Theorem 3.4.6, if and only if k is between $n + 2$ and $2n^2 - 3n$.

---

[5]Here, non-degenerate means that the curve is not contained in any abelian subvariety. Although this differs from our assumption on C, one of the first steps in the proof of this result is to establish that the maps P_I used in the proof of Lemma 3.4.4 contain open subsets of A^n in their images. Consequently, if C is non-degenerate, then the $\mathscr{X}$-dual web $\mathcal{W}_C^{\mathscr{X}}$ is generically smooth.

Chapter 4
The Converse to Abel's Theorem

The main result of this chapter is a converse to Abel's addition Theorem stated in Sect. 4.1. It ensures the algebraicity of local datum satisfying the hypotheses of Abel's addition Theorem. Its first version was established by Sophus Lie in the context of double-translation surfaces. Lie's arguments consisted in a tour-de-force analysis of an overdetermined system of PDEs. Later Poincaré introduced a geometrical method to handle the problem solved analytically by Lie. Poincaré's approach was later revisited, and made more precise by Darboux, to whom the approach presented in Sect. 4.2 can be traced back. By the way, those willing to take for granted the validity of the converse of Abel's Theorem can safely skip Sect. 4.2.

Blaschke's school reinterpreted the converse to Abel's Theorem in the language of web geometry to obtain the algebraicity of germs of linear webs admitting complete abelian relations. This result turned out to be a ubiquitous tool for the algebraization problem of germs of webs. This dual version also provides a complete description of the space of abelian relations of algebraic webs. This will be treated in Sect. 4.1.2.

Web geometry also owes Poincaré a method to algebraize a smooth, non-necessarily linear, maximal rank $2n$-web $\mathcal{W}$ on $(\mathbb{C}^n, 0)$. The strategy is based on the study of a certain natural map from $(\mathbb{C}^2, 0)$ to $\mathbb{P}(\mathcal{A}(\mathcal{W}))$, here called Poincaré's map of $\mathcal{W}$. Closely related are the canonical maps of the web. Their definition mimics the one of the canonical map of a projective curve. All this will be made precise in Sect. 4.3

Poincaré's original motivation had not much to do with web geometry, but focused instead on the relations between double-translation hypersurfaces and Theta divisors on Jacobian varieties of projective curves. In Sect. 4.4 these relations will be reviewed. Modern proofs of some of the theorems by Lie, Poincaré, and Wirtinger on the subject are also sketched.

© Springer International Publishing Switzerland 2015
J.V. Pereira, L. Pirio, *An Invitation to Web Geometry*, IMPA Monographs 2,
DOI 10.1007/978-3-319-14562-4_4

4.1   The Converse to Abel's Theorem

4.1.1   Statement

Let $H_0 \subset \mathbb{P}^n$, $n \geq 2$, be a hyperplane, and $k \geq 3$ be an integer. For $i \in \underline{k}$, let C_i be a germ of complex curve in $\mathbb{P}^n$ that intersects H_0 transversely at a point denoted by $p_i(H_0)$. Assume that the points $p_i(H_0)$ are pairwise distinct. Let also $p_i : (\check{\mathbb{P}}^n, H_0) \to C_i$ be the germs of holomorphic maps characterized by $H \cap C_i = \{p_i(H)\}$ for every $i \in \underline{k}$ and every $H \in (\check{\mathbb{P}}^n, H_0)$. For a picture, see Fig. 4.1.

Using the notation settled above, the converse to Abel's Theorem can be phrased as follows.

Theorem 4.1.1. *For $i \in \underline{k}$, let ω_i be a germ of non-zero holomorphic 1-form on C_i. If*

$$p_1^*(\omega_1) + \cdots + p_d^*(\omega_d) = 0 \qquad (4.1)$$

as a germ of 1-form on $(\check{\mathbb{P}}^n, H_0)$, then there exist an algebraic curve $C \subset \mathbb{P}^n$ of degree k, and an abelian differential $\omega \in H^0(C, \omega_C)$ such that $C_i \subset C$ and $\omega|_{C_i} = \omega_i$ for all $i \in \underline{k}$.

Starting from the middle 1970s, a number of generalizations of this remarkable result appeared in print. It has been generalized from curves to higher dimensional varieties in [62]; it has been shown in [79] that it suffices to have rational trace (4.1)

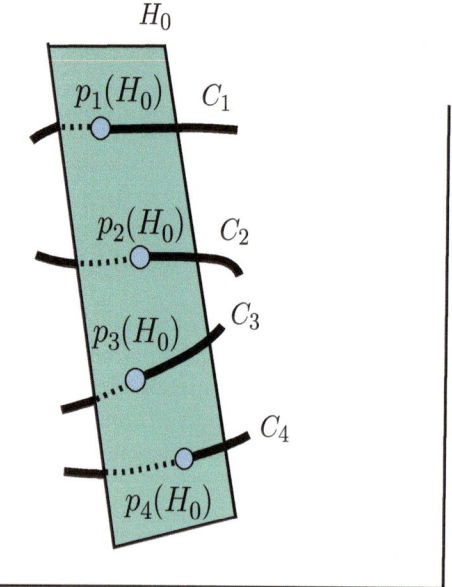

Fig. 4.1 Germs of analytic curves ...

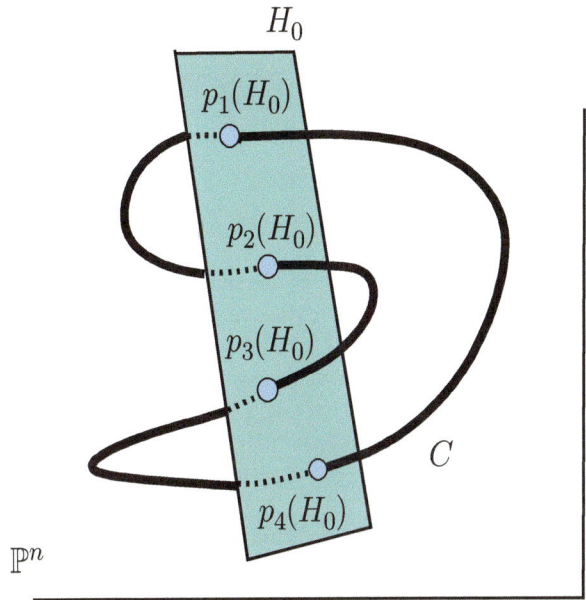

Fig. 4.2 ...globalize in the presence of a complete abelian relation (4.1)

to ensure the algebraicity of the data; and more general traces than the one with respect to hyperplanes have been considered in [50, 133] (Fig. 4.2).

The proof of Theorem 4.1.1 is postponed to Sect. 4.2. Assuming its validity, a dual formulation in terms of webs is given and proved below.

4.1.2 Dual Formulation

Let H_0 be a hyperplane in $\mathbb{P}^n$, $n \geq 2$. In Sect. 1.1.3 of Chap. 1 it was shown the existence of an equivalence between linear quasi-smooth k-webs on $(\check{\mathbb{P}}^n, H_0)$, and k germs of curves in $\mathbb{P}^n$ intersecting H_0 transversely in k distinct points. This equivalence implies the following variant of the converse to Abel's Theorem.

Theorem 4.1.2. *Let $\mathcal{W}$ be a linear quasi-smooth web on $(\check{\mathbb{P}}^n, H_0)$. If it admits a complete abelian relation, then it is algebraic. Accordingly, there exists a projective curve $C \subset \mathbb{P}^n$ transverse to H_0 such that $\mathcal{W}$ coincides with $\mathcal{W}_C(H_0)$. Furthermore, the space $\mathcal{A}(\mathcal{W})$ of abelian relations of $\mathcal{W}$ is naturally isomorphic to $H^0(C, \omega_C)$.*

Proof. For the reader's convenience, the proof starts by detailing the equivalence between germs of linear webs and germs of curves. Let $\mathcal{W} = \mathcal{F}_1 \boxtimes \cdots \boxtimes \mathcal{F}_k$ be a quasi-smooth linear k-web on $(\check{\mathbb{P}}^n, H_0)$. For each of the foliations $\mathcal{F}_i$, consider its Gauss map

$$\mathcal{G}_i : (\check{\mathbb{P}}^n, H_0) \longrightarrow \mathbb{P}^n$$

$$H \longmapsto T_H \mathcal{F}_i .$$

Since the foliation $\mathcal{F}_i$ has linear leaves, the map $\mathcal{G}_i$ is constant along them. Notice also that the restriction of $\mathcal{G}_i$ to any line transversal to the leaf of $\mathcal{F}_i$ passing through $H_0 \in \check{\mathbb{P}}^n$ is injective. These two facts taken together imply that $\mathcal{G}_i$ is a submersion defining $\mathcal{F}_i$ and that its image is a germ of smooth curve $C_{\mathcal{F}_i} \subset \mathbb{P}^n$ intersecting H_0 transversely (the latter being now viewed as a hyperplane in $\mathbb{P}^n$).

Notice that $\mathcal{G}_i$ associates with a hyperplane $H \in (\check{\mathbb{P}}^n, H_0)$ the intersection of $H \subset \mathbb{P}^n$ with $C_{\mathcal{F}_i}$. In other words, $\mathcal{G}_i = p_i$ in the notation used in the converse to Abel's Theorem.

If η_i is a germ of closed 1-form defining $\mathcal{F}_i$, then there exists a germ of holomorphic 1-form ω_i in $C_{\mathcal{F}_i}$ such that $\eta_i = \mathcal{G}_i^* \omega_i$. Therefore, if $\eta = (\eta_1, \ldots, \eta_k) \in \mathcal{A}(\mathcal{W})$ then there exist 1-forms $\omega_i \in \Omega^1(C_{\mathcal{F}_i}, \mathcal{G}_i(H_0))$ satisfying

$$\sum_{i=1}^{k} \mathcal{G}_i^*(\omega_i) = 0 .$$

When η is complete, none of the 1-forms ω_i vanishes identically. Thus the converse to Abel's Theorem ensures the existence of a projective curve $C \subset \mathbb{P}^n$ of degree k containing all the curves $C_{\mathcal{F}_i}$, and of an abelian differential $\omega \in H^0(C, \omega_C)$ which pull-backs through $\mathcal{G}_i$ to the ith component η_i of the abelian relation η. It is then clear that $\mathcal{W}$ is nothing but the germ of the algebraic web $\mathcal{W}_C$ at H_0. □

Corollary 4.1.3. *Let $\mathcal{W}$ be a linear smooth k-web on $(\mathbb{C}^n, 0)$. If $\mathcal{W}$ has maximal rank, then it is algebraic.*

Proof. It suffices to show the existence of a complete abelian relation. If there is none, then there exists a proper k'-subweb $\mathcal{W}'$ of $\mathcal{W}$ such that $\mathrm{rank}(\mathcal{W}') = \mathrm{rank}(\mathcal{W}) = \pi(n, k)$. But $\mathrm{rank}(\mathcal{W}') \leq \pi(n, k') < \pi(n, k)$ since $k' < k$. This contradiction proves the corollary. □

Corollary 4.1.4. *A smooth web of maximal rank is algebraizable if and only if it is linearizable.*

The latter corollary indicates the general strategy for solving the algebraization problem for webs

in order to prove that a web of maximal rank is algebraizable, it suffices to show that it is linearizable.

In fact, a similar strategy also applies to webs of higher codimension. Most of the known algebraization results in web geometry are proved in this way. The simplest instance of this approach will be the subject of Sect. 4.3. A considerably more involved instance will occupy the whole Chap. 5.

4.2 Proof of Theorem 4.1.1

The notation introduced in Sect. 4.1.1 is valid throughout this section.

4.2.1 Reduction to Dimension Two

This subsection aims to prove that the converse to Abel's Theorem in dimension n follows from the case of dimension $n - 1$ when $n > 2$.

Let us assume that $n > 2$ and consider a generic point $p \in H_0 \subset \mathbb{P}^n$. The linear projection $\pi : \mathbb{P}^n \dashrightarrow \mathbb{P}^{n-1}$ with center at p when restricted to the germs of curves C_i induces germs of biholomorphisms onto their images, which are denoted D_i. Moreover, (the germification at $\pi(H_0)$ of) the dual inclusion $\check{\pi} : \check{\mathbb{P}}^{n-1} \to \check{\mathbb{P}}^n$ fits into the following commutative diagram :

$$C_i \hookrightarrow \mathbb{P}^n \dashrightarrow{\ \pi\ } \mathbb{P}^{n-1} \hookleftarrow D_i$$

$$p_i \searrow \qquad \nearrow q_i = \pi \circ p_i \circ \check{\pi}$$

$$(\check{\mathbb{P}}^n, H_0) \xleftarrow{\ \check{\pi}\ } (\check{\mathbb{P}}^{n-1}, \pi(H_0)).$$

Since the restriction of π to C_i is a (germ of) biholomorphism onto D_i, the 1-form ω_i can be seen as a (germ of) 1-form on D_i. Under this identification, it is clear that

$$\sum_{i=1}^k q_i^* \omega_i = \check{\pi}^* \left(\sum_{i=1}^k p_i^* \omega_i \right) = 0.$$

The converse to Abel's Theorem in dimension $n - 1$ implies the existence of an algebraic curve $D \subset \mathbb{P}^{n-1}$ of degree k containing all the curves D_i, and of an abelian, thus rational, 1-form ω on D such that $\omega|_{D_i} = \omega_i$ for every $i \in \underline{k}$.

The closure $S = \overline{\pi^{-1}(D)}$ is the cone over D with vertex at p. Notice that it contains all the curves C_i and has dimension two.

Let $p' \in \mathbb{P}^n$ be another generic point of H_0. The same argument as above implies the existence of another surface S' containing the curves C_i. It follows that the curves C_i are contained in the intersection $S \cap S'$. Since S and S' are distinct (these are cones with distinct vertex in $\mathbb{P}^n$ and $n > 2$ by hypothesis), this suffices to ensure the existence of a projective curve C in $\mathbb{P}^n$ containing all the curves C_i. Furthermore, the pull-back by π of ω from D to C is a rational 1-form satisfying $\omega|_{C_i} = \omega$ for every $i \in \underline{k}$.

Thus, to establish Theorem 4.1.1 it suffices to consider the two-dimensional case. This will be done starting from the next section.

4.2.2 The Two-Dimensional Case: Preliminaries

To keep in mind that the ambient space has dimension 2, the hyperplanes in the statement of Theorem 4.1.1 will be denoted by ℓ_0 and ℓ, instead of H_0 and H.

Assume that the main hypothesis of the converse to Abel's Theorem is satisfied: for $i \in \underline{k}$, there are non-identically zero germs of holomorphic differentials $\omega_i \in \Omega_{C_i}^1$ such that (4.1) holds (Fig. 4.3). Let (x, y) be an affine system of coordinates on an affine chart $\mathbb{C}^2 \subset \mathbb{P}^2$ such that: (i) $\ell_0 \cap \mathbb{C}^2 = \{x = 0\}$; and (ii) none of the points $p_i(H_0)$ belong to $\ell_\infty = \mathbb{P}^2 \setminus \mathbb{C}^2$.

Since a generic line in the projective plane admits a unique affine equation of the form $x = ay + b$, the variables a and b can be considered as affine coordinates on $(\check{\mathbb{P}}^2, \ell_0)$. If $\ell_{a,b}$ denotes the line in $\mathbb{P}^2$ of affine equation $x = ay + b$, then $p_i(\ell_{a,b})$ can be written as $\big(x_i(a, b), y_i(a, b)\big)$, where $x_i, y_i : (\check{\mathbb{P}}^2, \ell_0) \to \mathbb{C}$ are two germs of holomorphic functions satisfying $x_i(a, b) = a\, y_i(a, b) + b$ identically on $(\check{\mathbb{P}}^2, \ell_0)$.

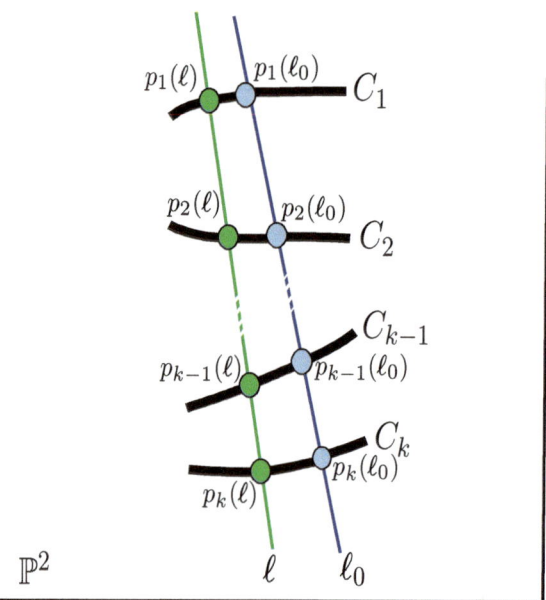

Fig. 4.3 Geometry of the converse to Abel's Theorem in $\mathbb{P}^2$

It will be convenient to assume that for every $i \in \underline{k}$, the function y_i is non-constant. Of course, this holds true for a generic choice of affine coordinates (x, y) on $\mathbb{C}^2 \subset \mathbb{P}^2$.

Let also $\eta_i \in \Omega^1(\check{\mathbb{P}}^2, \ell_0)$ be the pull-back of ω_i by p_i and $u_i : (\check{\mathbb{P}}^2, \ell_0) \to (\mathbb{C}, 0)$ be a primitive of η_i with value at ℓ_0 equal to zero, that is

$$\eta_i = p_i^* \omega_i \qquad \text{and} \qquad u_i(a, b) = u_i(\ell_{a,b}) = \int_{(0,0)}^{(a,b)} \eta_i, \quad (a, b) \in \mathbb{C}^2.$$

Differential Identities

One of the key ingredients of the proof of the converse to Abel's Theorem presented here is the following observation. It was first made in this context by Darboux.

Lemma 4.2.1. *For every $i \in \underline{k}$, the following differential equations are identically satisfied on $(\check{\mathbb{P}}^2, \ell_0)$:*

$$\frac{\partial y_i}{\partial a} = y_i \frac{\partial y_i}{\partial b}, \qquad \frac{\partial x_i}{\partial a} = y_i \frac{\partial x_i}{\partial b} \qquad \text{and} \qquad \frac{\partial u_i}{\partial a} = y_i \frac{\partial u_i}{\partial b}. \tag{4.2}$$

Proof. First notice that, for a fixed $i \in \underline{k}$, the functions x_i, y_i, and u_i define the very same foliation on $(\check{\mathbb{P}}^2, \ell_0)$. Consequently, the 1-forms dx_i, dy_i, and du_i are all proportional. Thus, it suffices to prove the identity

$$\frac{\partial y_i}{\partial a} = y_i \frac{\partial y_i}{\partial b} \tag{4.3}$$

to obtain the other two.

Since C_i intersects $\ell_0 = \{x = 0\}$ transversely at $(0, y_0)$, there exists a germ $g_i \in \mathcal{O}_{(\mathbb{C},0)}$ for which $C_i = \{y = g_i(x)\}$. Therefore, for all $(a, b) \in (\check{\mathbb{P}}^2, \ell_0)$, one has

$$y_i(a, b) = g_i(x_i(a, b)) = g_i(a \, y_i(a, b) + b).$$

Differentiation of this identity implies

$$\frac{\partial y_i}{\partial a}\left(1 - a \, g_i'(x_i)\right) = y_i \, g_i'(x_i) \qquad \text{and} \qquad \frac{\partial y_i}{\partial b}\left(1 - a \, g_i'(x_i)\right) = g_i'(x_i).$$

By considering the second differential identity just above, it follows that the function $(a, b) \mapsto 1 - a \, g_i'(x_i(a, b))$ does not vanish identically. Dividing by $1 - a \, g_i'(x_i(a, b))$ the two preceding identities, one obtains therefore that y_i verifies (4.3). The lemma follows. $\qquad\square$

Notice that the preceding lemma combined with the hypothesis $\sum_i p_i^* \omega_i = \sum_i du_i = 0$ gives us the following relations

$$\sum_{i=1}^{k} \frac{\partial u_i}{\partial a} = \sum_{i=1}^{k} y_i \frac{\partial u_i}{\partial b} \equiv 0 \quad \text{and} \quad \sum_{i=1}^{k} \frac{\partial u_i}{\partial b} \equiv 0. \tag{4.4}$$

4.2.3   Lifting to the Incidence Variety

Let $\mathcal{I} \subset \mathbb{P}^2 \times \check{\mathbb{P}}^2$ be the incidence variety, that is

$$\mathcal{I} = \left\{ (p, \ell) \in \mathbb{P}^2 \times \check{\mathbb{P}}^2 \mid p \in \ell \right\}.$$

As in Sect. 1.4.2 of Chap. 1, let $\pi : \mathcal{I} \to \mathbb{P}^2$ and $\check{\pi} : \mathcal{I} \to \check{\mathbb{P}}^2$ be the natural projections.

There is an open affine subset $V \subset \mathcal{I}$ isomorphic to the closed subvariety of $\mathbb{C}^2 \times \mathbb{C}^2$ defined by the equation $x = ay + b$, where (x, y) and (a, b) are, respectively, the affine coordinates on $\mathbb{C}^2 \subset \mathbb{P}^2$ and on $(\check{\mathbb{P}}^2, \ell_0) \subset \check{\mathbb{P}}^2$ used above.

Notice that V is isomorphic to $\mathbb{C}^3$ and that (y, a, b) is an affine coordinate system on it. Using these coordinates, define the germ of meromorphic 2-form $\check{\Psi}_0$ on $(\mathbb{C}^3, \{ay + b = 0\})$[1]

$$\check{\Psi}_0(a, b, y) = \sum_{i=1}^{d} \frac{\eta_i(a, b) \wedge dy}{y - y_i(a, b)}. \tag{4.5}$$

Recall that the 1-forms $\eta_i \in \Omega^1(\check{\mathbb{P}}^n, \ell_0)$ were introduced in Sect. 4.1.2 as the ones corresponding to the 1-forms ω_i via projective duality. Notice that $\check{\Psi}_0$ is the restriction on V of a germ of meromorphic 2-form Ψ on $\check{\pi}^{-1}(\check{\mathbb{P}}^2, \ell_0) = (\mathcal{I}, \check{\pi}^{-1}(\ell_0))$.

Remark 4.2.2. To understand the idea behind the definition of Ψ, imagine that the local datum $\{(C_i, \omega_i)\}$ is indeed the germification along ℓ_0 of a global curve C and of an abelian differential $\omega \in H^0(C, \omega_C)$. In this case, there exists a meromorphic 2-form Ω on $\mathbb{P}^2$ satisfying $\text{Res}_C \Omega = \omega$. Writing down the meromorphic 2-form $\pi^* \Omega$ in the coordinates (a, b, y), one ends up with an expression exactly like (4.5).

---

[1]Here and throughout in the proof of the converse of Abel's Theorem, the notation (X, Y) means **the germ of the variety** X **along** (or **at**) Y. One should think of open subsets of X, arbitrarily small among the ones containing Y.

Recall that the 1-form η_i is equal to du_i, and that u_i verifies (4.2). Therefore $\eta_i = (\partial u_i/\partial b)(y_i da + db)$ for every $i \in \underline{k}$. It is then easy to determine the expression $\Psi_0(x, y, a)$ of Ψ in the coordinates x, y, a. Using (4.4) one obtains

$$\Psi_0(x, y, a) = \sum_{i=1}^{k} \frac{\frac{\partial u_i}{\partial b}(a, x - ay)}{y - y_i(a, x - ay)} \, dx \wedge dy \, .$$

If F is the germ of meromorphic function on $\check{\pi}^{-1}(\check{\mathbb{P}}^2, \ell_0)$ which in the coordinates (x, y, a) can be written as

$$F_0(x, y, a) = \sum_{i=1}^{k} \frac{\frac{\partial u_i}{\partial b}(a, x - ay)}{y - y_i(a, x - ay)}$$

then

$$\Psi_0(x, y, a) = F_0(x, y, a) \, dx \wedge dy \, .$$

4.2.4 Back to the Projective Plane

The next step of the proof consists in showing that the 2-form Ψ defined above comes from a 2-form on the projective plane.

Lemma 4.2.3. *There exists a germ of meromorphic function f on $(\mathbb{P}^2, \ell_0)$ such that $F = \pi^*(f) = f \circ \pi$. Consequently, $\Psi = \pi^*\Omega$, where Ω is the meromorphic 2-form $f(x, y)dx \wedge dy$ on $(\mathbb{P}^2, \ell_0)$.*

Proof. It suffices to prove that $\frac{\partial F_0}{\partial a}$ is identically zero. For that sake, let $\check{F}_0$ be the expression for F in the coordinate system (a, b, y), that is,

$$\check{F}_0(a, b, y) = F_0(ay + b, y, a) = \sum_{i=1}^{k} \frac{\frac{\partial u_i}{\partial b}(a, b)}{y - y_i(a, b)} \, .$$

Differentiating the latter identity with respect to a and using (4.2), one obtains

$$\frac{\partial \check{F}_0}{\partial a} = \sum_{i=1}^{k} \frac{y_i'(\frac{\partial u_i}{\partial b})^2 + y_i \frac{\partial^2 u_i}{\partial b^2}}{y - y_i} + \sum_{i=1}^{k} \frac{y_i' y_i (\frac{\partial u_i}{\partial b})^2}{(y - y_i)^2} \tag{4.6}$$

with $y_i' = dy_i/du_i$ for every $i \in \underline{k}$.

On the other hand, since $\sum_i \frac{\partial u_i}{\partial b} = 0$ according to (4.4), one has

$$y\,\check{F}_0 = \sum_{i=1}^{k} \frac{y\,\frac{\partial u_i}{\partial b}}{y - y_i} = \sum_{i=1}^{k} \frac{\partial u_i}{\partial b} + \sum_{i=1}^{k} \frac{y_i\,\frac{\partial u_i}{\partial b}}{y - y_i} = \sum_{i=1}^{d} \frac{y_i\,\frac{\partial u_i}{\partial b}}{y - y_i}.$$

Differentiating this expression with respect to b gives the identity

$$y\frac{\partial \check{F}_0}{\partial b} = \sum_{i=1}^{k} \frac{y_i'(\frac{\partial u_i}{\partial b})^2 + y_i\,\frac{\partial^2 u_i}{\partial b^2}}{y - y_i} + \sum_{i=1}^{k} \frac{y_i'\,y_i\,(\frac{\partial u_i}{\partial b})^2}{(y - y_i)^2}. \qquad (4.7)$$

From (4.6) and (4.7), it follows that $\check{F}_0$ satisfies the partial differential equation

$$\frac{\partial \check{F}_0}{\partial a} - y\frac{\partial \check{F}_0}{\partial b} = 0.$$

To conclude, it suffices to notice that

$$\frac{\partial F_0}{\partial a}(x, y, a) = \left(\frac{\partial \check{F}_0}{\partial a} - y\frac{\partial \check{F}_0}{\partial b}\right)(a, x - ay, y) = 0.$$

$\square$

Recovering the Curves

Now, notice that the polar set $(\Omega)_\infty$ of the 2-form Ω is nothing more than the union of the germs C_i with $i \in \underline{k}$.

Lemma 4.2.4. *If Ω is the 2-form provided by Lemma 4.2.3, then*

$$(\Omega)_\infty = \bigcup_{i \in \underline{k}} C_i.$$

Proof. Consider $\Psi = \pi^*\Omega$ in the coordinates $a, b, z = 1/y$. Performing the change of variable $y = 1/z$ in (4.5) gives, for z close to 0:

$$\Psi(a, b, z) = -\sum_{i=1}^{k} \frac{\frac{\partial u_i}{\partial b}(y_i\,da + db)}{z(1 - z\,y_i)} \wedge dz$$

$$= -\sum_{r \geq 0}\sum_{i=1}^{k} \left[\left(\frac{\partial u_i}{\partial b} y_i^{r+1} z^{r-1}\right)da + \left(\frac{\partial u_i}{\partial b} y_i^r z^{r-1}\right)db\right] \wedge dz.$$

Lemma 4.2.1 implies the vanishing of the coefficients of $z^{-1}da$ and $z^{-1}db$. Thus $\Psi(a,b,z)$ is holomorphic at an open neighborhood of $\{z = 0\}$. Therefore Ω is holomorphic at a neighborhood of $\ell_0 \cap \ell_\infty$.

For $(a,b) \in (\check{\mathbb{P}}^2, \ell_0)$, the restriction of the function f given by Lemma 4.2.3 at the line $\ell_{a,b} \subset (\check{\mathbb{P}}^2, \ell_0)$ is

$$f_{a,b}(y) = \check{F}_0(a,b,y) = \sum_{i=1}^{k} \frac{\frac{\partial u_i}{\partial b}(a,b)}{y - y_i(a,b)}.$$

Recall that $du_i = \eta_i$. According to the hypotheses, the partial derivative $\partial u_i / \partial b$ does not vanish identically on C_i, for any $i \in \underline{k}$. Hence, for a generic $(a,b) \in (\check{\mathbb{P}}^2, \ell_0)$, one has

$$\mathbb{C}^2 \cap (f_{a,b})_\infty = p_1(a,b) + \cdots + p_k(a,b).$$

This suffices to prove the lemma. $\qquad\qquad\qquad\qquad\qquad\qquad\qquad\qquad$ □

Recovering the Differential Forms

It is also possible to extract from Ω the 1-forms $\omega_1, \ldots, \omega_k$ with the help of Poincaré's residue.

Lemma 4.2.5. *For every $i \in \underline{k}$, one has* $\mathrm{Res}_{C_i} \Omega = \omega_i$.

Proof. Fix $i \in \underline{k}$ and set $p = p_i(\ell_0) = (0, y_i(0,0)) \in C_i$. Let q be the point in $\pi^{-1}(\ell_0)$ which is represented by $(0, 0, y_i(0,0))$ in the coordinate system (a, b, y). If

$$(D, q) = \left\{ (a,b,y) \in \pi^{-1}\big((\check{\mathbb{P}}^2, \ell_0)\big) \,\big|\, a = 0 \right\}$$

then the restriction of π to (D, q) is a germ of biholomorphism

$$\rho : (D, q) \to (\check{\mathbb{P}}^2, p).$$

In the coordinates (b, y) on D,

(a) the pull-back of C_i to D is $E_i = \rho^{-1}(C_i) = \{y - y_i(0,b) = 0\}$;
(b) the 1-form $\rho^*(\omega_i)$ coincides with $\check{\pi}^*(\eta_i)|_{E_i}$; and
(c) the pull-back of Ω to (D, q) by ρ can be written as

$$\rho^*(\Omega) = \sum_{i=1}^{k} \frac{\eta_i(0, b) \wedge dy}{y - y_i(0, b)}.$$

Item (a) implies that $\frac{\eta_j \wedge dy}{y - y_j}$ is holomorphic in a neighborhood of E_i when $j \neq i$. Items (b) and (c), in their turn, imply

$$\mathrm{Res}_{E_i}\left(\rho^*\Omega\right) = \mathrm{Res}_{E_i}\left(\frac{\eta_i(0,b)}{y - y_i(0,b)} \wedge dy\right) = \eta_i(0,b)\Big|_{E_i} = \rho^*\omega_i \, .$$

Since ρ is an isomorphism, it follows that

$$\mathrm{Res}_{C_i}\left(\Omega\right) = \omega_i$$

for every $i \in \underline{k}$. The lemma is proved. $\square$

4.2.5 Globalizing to Conclude

At this point, to conclude the proof of Theorem 4.1.1, it suffices to prove that the 2-form Ω is the restriction on $(\mathbb{P}^2, \ell_0)$ of a rational 2-form on $\mathbb{P}^2$. Indeed, if this is the case, then the polar set of Ω will be a projective curve C containing the curves C_i, and its residue along C will be an abelian differential $\omega \in H^0(C, \omega_C)$ according to Proposition 3.2.10.

The globalization of Ω follows from a particular case of a classical result stated below as a lemma.

Lemma 4.2.6. *Let $\ell \subset \mathbb{P}^2$ be a line. Any germ of meromorphic function $g :$ $(\mathbb{P}^2, \ell) \dashrightarrow \mathbb{P}^1$ extends to a rational function on $\mathbb{P}^2$.*

Proof. Suppose that ℓ is the line at infinity and let (x, y) be affine coordinates on $\mathbb{C}^2 = \mathbb{P}^2 \setminus \ell$.

Fix an arbitrary representative of the germ g defined in a neighborhood U of the line at infinity. Still denote it by g. Notice that the restriction of U to $\mathbb{C}^2$ contains the complement of a polydisc Δ. Consider the Laurent expansion of g on $\mathbb{C}^2$:

$$g(x, y) = \sum_{i,j \in \mathbb{Z}^2} a_{ij} x^i y^j \, . \tag{4.8}$$

Since it converges in a neighborhood of infinity to a meromorphic function, in order to prove the lemma, it suffices to show that there exists $(i_0, j_0) \in \mathbb{Z}$ such that

$$\Gamma = \{(i, j) \in \mathbb{Z}^2 \mid a_{ij} \neq 0\} \subset (i_0, j_0) + \mathbb{N}^2 \, . \tag{4.9}$$

Indeed, in this case g extends meromorphically to $\mathbb{C}^2$ hence to the whole projective plane. And it is well known that a global meromorphic function on $\mathbb{P}^2$ is rational (this follows immediately from Chow's theorem, see [63, p. 168] for instance).

To prove that (4.9) holds true for some $i_0, j_0 \in \mathbb{Z}$, rewrite the Laurent series (4.8) as

$$g(x, y) = \sum_{i \in \mathbb{Z}} b_i(y)x^i \,,$$

and consider the function $I : \mathbb{C} \to \mathbb{Z} \cup \{-\infty\}$ defined by

$$I(t) = \inf \{ i \in \mathbb{Z} \,|\, b_i(t) \neq 0 \} \,.$$

Let us assume that g is not constant (in this case, the lemma is trivial). Because it is assumed to be meromorphic, the zero divisor of g along the "line at infinity" ℓ is a finite union of (germs of) analytic hypersurfaces on $(\mathbb{P}^2, \ell)$. From this, it follows that if $|t| > c$ for a certain positive constant $c \in \mathbb{R}$, then the function $x \mapsto g(x, t)$ is a non-trivial global meromorphic function of x. Therefore $I(t) \in \mathbb{Z}$ if $|t| > c$. Since $\mathbb{Z} \cup \{-\infty\}$ is a countable set, there exists an integer i_0, and an uncountable set $\Sigma \subset \mathbb{C}$ for which the restriction of b_i to Σ is zero whenever $i \leq i_0$. Therefore, for $i \leq i_0$, the functions b_i are indeed zero all over $\mathbb{C}$. In other words, $\Gamma \subset (i_0 + \mathbb{N}) \times \mathbb{Z}$. As explained above, the lemma follows. □

4.3 Algebraization of Smooth $2n$-Webs

As the title of this chapter indicates, Theorem 4.1.2 is a ubiquitous tool when the algebraization of webs comes to mind. It is worth repeating that most of the known algebraization results use the abelian relations in order to linearize the web and then apply the converse to Abel's Theorem in its dual formulation.

As promised, the simplest instance where this strategy applies—the case of smooth $2n$-webs $\mathcal{W}$ on $(\mathbb{C}^{2n}, 0)$ of maximal rank—is treated below.

4.3.1 Poincaré's Map

Let

$$\mathcal{W} = \mathcal{F}_1 \boxtimes \cdots \boxtimes \mathcal{F}_{2n} = \mathcal{W}(\omega_1, \ldots, \omega_{2n})$$

be a smooth $2n$-web on $(\mathbb{C}^n, 0)$. Assume that $\mathcal{W}$ has maximal rank. Since $\pi(n, 2n) = n + 1$, the space $\mathcal{A}(\mathcal{W})$ is a complex vector space of dimension $n + 1$. In particular, according to Corollary 2.3.11, one has

$$\dim F^1 \mathcal{A}(\mathcal{W}) = 1 \,.$$

If $F_x^\bullet \mathcal{A}(\mathcal{W})$ denotes the corresponding filtration of $\mathcal{A}(\mathcal{W})$ centered[2] at $x \in$ $(\mathbb{C}^n, 0)$, then one still has dim $F_x^1 \mathcal{A}(\mathcal{W}) = 1$. **Poincaré's map** of $\mathcal{W}$ is defined as

$$P_{\mathcal{W}} : (\mathbb{C}^n, 0) \longrightarrow \mathbb{P}\mathcal{A}(W)$$

$$x \longmapsto \left[F_x^1 \mathcal{A}(\mathcal{W}) \right].$$

It is a covariant of $\mathcal{W}$: if $\varphi \in \mathrm{Diff}(\mathbb{C}^n, 0)$ then

$$P_{\varphi * \mathcal{W}} = \varphi^* \left(P_{\mathcal{W}} \right) = P_{\mathcal{W}} \circ \varphi.$$

4.3.2 Linearization

For every $i \in \underline{2n}$ and each $x \in (\mathbb{C}^n, 0)$, consider the linear map

$$ev_i(x) : \mathcal{A}(\mathcal{W}) \longrightarrow \Omega^1(\mathbb{C}^n, x)$$

$$\eta = (\eta_j)_{j=1}^{2n} \longmapsto \eta_i(x).$$

The kernel of $ev_i(x)$ corresponds to the abelian relations η of $\mathcal{W}$ whose i-component η_i vanishes at x.

Lemma 4.3.1. *For every $i \in \underline{2n}$, the linear map $ev_i(x)$ has rank equal to one. In particular, the family of kernels $\{\ker(ev_i(x))\}_{x \in (\mathbb{C}^n, 0)}$ is a subbundle of corank one of the trivial bundle over $(\mathbb{C}^n, 0)$ with fiber $\mathcal{A}(\mathcal{W})$. Moreover, for every subset $I \subset \underline{2n}$ of cardinality n, the following identity holds true*

$$\bigcap_{i \in I} \ker ev_i(x) = F_x^1 \mathcal{A}(\mathcal{W}).$$

Proof. If $ev_i(x)(\eta) \neq 0$, the 1-form η_i defines the foliation $\mathcal{F}_i$ in a neighborhood of x. This shows that the rank of $ev_i(x)$ is at most one. By semi-continuity, if it is not equal to 1, then it must be equal to 0 on $(\mathbb{C}^n, 0)$. In other words, the ith component of every abelian relation $\eta \in \mathcal{A}(\mathcal{W})$ vanishes at the origin. Therefore

$$\dim \frac{F^0 \mathcal{A}(\mathcal{W})}{F^1 \mathcal{A}(\mathcal{W})} \leq (2n - 1) - \ell^1(\mathcal{W}) = n - 1$$

---

[2]Here and throughout, the convention about germs made in Sect. 1.1.1 is in use. If one wants to be more precise, then $\mathcal{W}$ has to be thought as a web defined on an open subset U of $\mathbb{C}^n$ containing the origin and $F_x^1 \mathcal{A}(\mathcal{W})$ is the filtration of the germ of $\mathcal{W}$ at x.

according to the proof of Lemma 2.3.6. But then, according to Corollary 2.3.11, one has $\mathrm{rank}(\mathcal{W}) \leq (n-1) + 1 = n < \pi(n, 2n)$ contradicting the maximality hypothesis on the rank.

To prove the second part of the lemma, notice that the smoothness of $\mathcal{W}$ ensures the linear independence of $T_x \mathcal{F}_i$ with $i \in \underline{2n} \setminus I$. □

Proposition 4.3.2. *If L is a leaf of $\mathcal{W}$ then $P_{\mathcal{W}}(L)$, its image under Poincaré's map, is contained in a hyperplane.*

Proof. From Lemma 4.3.1, it follows that for every $i \in \underline{2n}$ and every $x \in (\mathbb{C}^n, 0)$, $P_{\mathcal{W}}(x) = \mathbb{P}F_x^1 \mathcal{A}(\mathcal{W})$ is contained in $\mathbb{P} \ker ev_i(x)$ that is a hyperplane in $\mathbb{P}\mathcal{A}(\mathcal{W})$.

Fix $i \in \underline{2n}$ and let $u_i : (\mathbb{C}^n, 0) \to \mathbb{C}$ be a submersion defining $\mathcal{F}_i$. A leaf L of $\mathcal{F}_i$ is a level hypersurface of u_i. The i-component η_i of an abelian relation $\eta \in \mathcal{A}(\mathcal{W})$ is of the form $g(u_i)du_i$. Therefore if $x, y \in L$ are two points of L, then η_i vanishes at x if and only it vanishes at y. Thus $\ker ev_i(x) = \ker ev_i(y)$ for every $x, y \in L$. Hence, the image of L by $P_{\mathcal{W}}$ is contained in the hyperplane $\mathbb{P} \ker ev_i(x) \subset \mathbb{P}\mathcal{A}(W)$ determined by any $x \in L$. □

Proposition 4.3.3. *Poincaré's map $P_{\mathcal{W}} : (\mathbb{C}^n, 0) \to \mathbb{P}\mathcal{A}(\mathcal{W})$ is a germ of biholomorphism.*

Proof. For $i \in \underline{n}$, let L_i be the leaf of $\mathcal{F}_i$ through zero. If $u_i : (\mathbb{C}^n, 0) \to (\mathbb{C}, 0)$ is a submersion defining $\mathcal{F}_i$, then $L_i = u_i^{-1}(0)$. Because $\mathcal{W}$ is smooth,

$$C_i = \bigcap_{j \in \underline{n} \setminus \{i\}} L_j$$

is a smooth germ of curve in $(\mathbb{C}^n, 0)$.

If $\gamma_i : (\mathbb{C}, 0) \to C_i$ is a regular parametrization of C_i, then it follows from (the proof of) Proposition 4.3.2 above that the image of $P_{\mathcal{W}} \circ \gamma_i$ is contained in

$$\ell_i = \bigcap_{j \in \underline{n} \setminus \{i\}} \mathbb{P} \ker ev_i(0)$$

which is a line in $\mathbb{P}\mathcal{A}(\mathcal{W})$ according to Lemma 4.3.1.

Since the tangent spaces of the lines $\ell_1, \ldots, \ell_n$ at $P_{\mathcal{W}}(0)$ generate $T_{P_{\mathcal{W}}(0)}\mathbb{P}\mathcal{A}(\mathcal{W})$, it suffices to show that $\Gamma_i = P_{\mathcal{W}} \circ \gamma_i : (\mathbb{C}, 0) \to \ell_i \subset \mathbb{P}^n$ has non-zero derivative at $0 \in \mathbb{C}$, for every $i \in \underline{n}$. But Γ_i is a map between germs of smooth curves, hence everything boils down to the injectivity of Γ_i for a fixed $i \in \underline{n}$.

If Γ_i is not injective, then there are pairs of distinct points $x, y \in C_i$ arbitrarily close to zero such that $\ker ev_i(x) = \ker ev_i(y)$. Hence, if the ith component of an abelian relation of $\mathcal{W}$ vanishes at x, then it also vanishes at y. It follows that the ith component of the abelian relation generating $F_0^1 \mathcal{A}(W)$ vanishes at the origin with multiplicity two. But this contradicts the equality $\ell^2(\mathcal{W}) = 2n - 1$ established in Proposition 2.3.1. Thus Γ_i is injective for any $i \in \underline{n}$. Consequently, the differential of $P_{\mathcal{W}}$ at the origin is invertible. The proposition follows. $\square$

4.3.3   Algebraization

It is a simple matter to put the previous results together in order to prove the following algebraization result.

Theorem 4.3.4. *A smooth $2n$-web $\mathcal{W}$ on $(\mathbb{C}^n, 0)$ of maximal rank is algebraizable. More precisely, its push-forward by its Poincaré's map is a web $\mathcal{W}_C$ where $C \subset \mathbb{P}^n$ is a $\mathcal{W}$-generic projective curve of degree $2n$ and genus $n + 1$.*

Proof. According to Proposition 4.3.3, $P_{\mathcal{W}}$ is a germ of biholomorphism. Hence $(P_{\mathcal{W}})_*(\mathcal{W})$ is a smooth $2n$-web equivalent to $\mathcal{W}$. In particular, its rank is also maximal. Proposition 4.3.2 implies that $(P_{\mathcal{W}})_*(\mathcal{W})$ is a linear web. Applying Corollary 4.1.3, one gets that the latter is algebraic. $\square$

4.3.4   Poincaré's Map for Planar Webs

It is possible to generalize Poincaré's map for smooth k-webs on $(\mathbb{C}^n, 0)$ for all integers n and k such that $2n \leq k$. The idea is to consider the last possibly non-trivial piece $F^l \mathcal{A}(\mathcal{W}) \neq 0$ of the filtration $F^\bullet \mathcal{A}(\mathcal{W})$. To be more precise, let us set, as in Remark 2.3.10,

$$\rho = \left\lfloor \frac{k - n - 1}{n - 1} \right\rfloor \qquad \text{and} \qquad \epsilon = (k - n - 1) - \rho(n - 1) \in \{0, \dots, n - 2\}.$$

With this notation, the last possibly non-trivial piece of $F^\bullet \mathcal{A}(\mathcal{W})$ is $F^{\rho+1} \mathcal{A}(\mathcal{W})$. Notice that $F^{\rho+1} \mathcal{A}(\mathcal{W})$ is not trivial indeed when $\mathcal{W}$ has maximal rank.

Then set $e = \dim F^{\rho+1} \mathcal{A}(\mathcal{W}) = \epsilon + 1 > 0$ and define Poincaré's map of $\mathcal{W}$ as

$$P_{\mathcal{W}} : (\mathbb{C}^n, 0) \longrightarrow \mathrm{Grass}\big(\mathcal{A}(W), e\big)$$

$$x \longmapsto F_x^{\rho+1} \mathcal{A}(\mathcal{W}).$$

When $e = 1$, that is $n = 2$ or $k \equiv 2 \mod (n - 1)$, then one still gets a map to a projective space with remarkable properties as shown in the next result for $n = 2$. Nevertheless, it does not linearize the web as when $k = 2n$.

Proposition 4.3.5. *If* W *is a smooth* k-*web of maximal rank on* $(\mathbb{C}^2, 0)$, *then* P_W
is an immersion. Moreover if S_W *is the image of* P_W *and if* L *is a leaf of* W *then
the following assertions hold:*

(a) *the image of* L *by* P_W *is contained in a projective space of codimension* $k - 3$;
(b) *the union of the projective tangent spaces of the surface* S_W *along the points of
the image of* L, *that is*

$$\bigcup_{x \in P_W(L)} T_x S_W,$$

is contained in a projective space of codimension $k - 4$;
(c) *more generally, for any integer* $s \leq k - 4$, *the union of the projective osculating
spaces of* S_W *of order* s *along the points of the image of* L, *that is*

$$\bigcup_{x \in P_W(L)} T_x^{(s)} S_W,$$

is contained in a projective space of codimension $k - (3 + s)$.

Proof. The proof is the natural generalization of the arguments used to prove
Propositions 4.3.2 and 4.3.3. The reader is invited to fill in the details of the
arguments sketched below.

Instead of considering the evaluation morphism $ev_i^{(0)}(x) = ev_i(x) : \mathcal{A}(W) \to \mathbb{C}$
one has to consider the higher order evaluation morphism $ev_i^{(j)}(x) : \mathcal{A}(W) \to$
$\mathbb{C}^{j+1}$ which associates with an abelian relation of W the jth jet at x of its
ith component. More explicitly, if u_i is a submersion defining $\mathcal{F}_i$ and if the ith
component of $\eta \in \mathcal{A}(W)$ is $\eta_i = f(u_i)du_i$, then

$$ev_i^{(j)}(x)(\eta) = f\big(u_i(x)\big) + tf'\big(u_i(x)\big) + \cdots + t^j f^{(j)}\big(u_i(x)\big)$$

where $(1, t, \ldots, t^j)$ is seen as a basis of $\mathbb{C}^{j+1}$.

It is a simple matter to adapt the arguments used to prove Proposition 4.3.3 in
order to show that P_W is an immersion.

To prove item (a), notice that $ev_i^{(k-4)}(x)$ has maximal rank. Consequently
$\dim \ker ev_i^{(k-4)}(x) = \pi(2, k) - (k - 3)$. Furthermore, if L is the leaf of $\mathcal{F}_i$ through
x then its image $P_W(L)$ is contained in $\mathbb{P} \ker ev_i^{(k-4)}(x)$. Put together, these two
remarks imply item (a).

To prove items (b) and (c), the key point is to notice the following. If $f :$
$(\mathbb{C}^2, 0) \to \mathcal{A}(W)$ is such that $f(x) \in \ker ev_i^{(k-4)}(x)$ for every $x \in (\mathbb{C}^2, 0)$, then the
derivatives of f at x with order at most s lie in $\ker ev_i^{(k-4-s)}(x)$.   □

The discussion of Poincaré's map for planar webs of maximal rank just made
leads naturally to the following characterization of algebraizable planar webs.

Theorem 4.3.6. *Let W is a smooth planar k-web of maximal rank with $k \geq 4$. The two following assertions are equivalent:*

(a) the image of Poincaré's map P_W is contained in the $(k-3)$th Veronese surface;
(b) the web W is algebraizable.

Proof. Let $S = S_W$ be the image of P_W. Suppose first that S is contained in the $(k-3)$th Veronese surface and let $\nu = \nu_{k-3} : \mathbb{P}^2 \to \mathbb{P}\mathcal{A}(W)$ be the corresponding Veronese embedding. According to Proposition 4.3.5, for every $i \in \underline{k}$ and every $x \in (\mathbb{C}^2, 0)$, the image of L (the leaf of $\mathcal{F}_i$ through x) is contained in a hyperplane H that osculates S hence $\nu(\mathbb{P}^2)$ up to order $k-3$ along $P_W(L)$. But this means that $\nu^*(H \cdot \nu(\mathbb{P}^2))$ is a curve in $\mathbb{P}^2$ with an irreducible component C for which every point is a singularity with algebraic multiplicity $k-3$. Since $\nu^* H$ has degree $(k-3)$, it follows that $C = (k-3)\ell$ for some line $\ell \subset \mathbb{P}^2$. This implies in particular that $\nu^{-1} \circ P_W(L) \subset \ell$. Thus the composition $\nu^{-1} \circ P_W$ linearizes W. Corollary 4.1.3 implies that $(\nu^{-1} \circ P_W)_* W$ is an algebraic web.

Conversely, if $W = W_C(H_0)$ is algebraic, then it is a simple matter to show that $P_W : (\check{\mathbb{P}}^2, H_0) \to \mathbb{P}\mathcal{A}(W)$ is the germ at H_0 of the $(k-3)$th Veronese embedding. For details see, for instance, [107].   □

Notice that when $k = 4$, assertion (a) is always fulfilled and one recovers Lie-Poincaré's theorem (that is Theorem 4.3.4 when $n = 2$).

For $k = 5$ the corresponding statement appears in [17, p. 255], and the proof there presented involves the study of Poincaré Blaschke's map of the web, which will be introduced in Chap. 5. The result for arbitrary k as stated above was proved in [73] using arguments very similar to the ones in the book just mentioned. The proof above uses slightly different arguments.

4.3.5   Canonical Maps

For a k-web W on $(\mathbb{C}^n, 0)$ of positive rank, it is possible to mimic the construction of canonical maps for projective curves as follows. Notice that the evaluation morphisms $ev_i(x)$ are linear functionals on $\mathcal{A}(W)$. As such, when non-zero, they can be seen as points of $\mathcal{A}(W)^*$.

For every $i \in \underline{k}$ such that ev_i does not vanish identically,[3] one considers the germ of meromorphic map

$$\kappa_{W,i} : (\mathbb{C}^n, 0) \dashrightarrow \mathbb{P}\mathcal{A}(W)^*$$

$$x \longmapsto \left[ev_i(x)\right].$$

By definition, $\kappa_{W,i}$ is the i**th canonical map** of W.

---

[3]This is the case if and only if W has an abelian relation whose ith component is non-trivial.

Lemma 4.3.1, or more precisely its proof, shows that when $\mathcal{W}$ is smooth and of maximal rank, $\kappa_{\mathcal{W},i}$ is defined for every $i \in \underline{k}$ and is regular at the origin of $\mathbb{C}^n$. Moreover, if $\mathcal{W} = \mathcal{W}_C(H_0)$ is an algebraic web on $(\check{\mathbb{P}}^n, H_0)$ dual to a projective curve $C \subset \mathbb{P}^n$ then, after identifying $H^0(C, \omega_C)$ with $\mathcal{A}(\mathcal{W})$ (*cf.* Theorem 3.2.8), one can put together any one of the canonical maps of $\mathcal{W}$ with the canonical map[4] of C in the following commutative diagram:

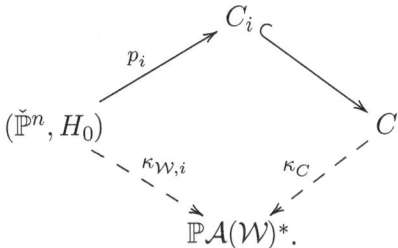

For smooth $2n$-webs on $(\mathbb{C}^n, 0)$ of maximal rank, Poincaré's map and the canonical maps of $\mathcal{W}$ are related through the formula

$$P_{\mathcal{W}}(x) = \bigcap_{i \in \underline{2n}} \kappa_{\mathcal{W},i}(x),$$

where the points $\kappa_{\mathcal{W},i}(x)$ are interpreted as hyperplanes in $\mathbb{P}\mathcal{A}(\mathcal{W})$.

Notice that when they are defined and for no matter which k and no matter which rank, the canonical maps always take values in the dual projective space $\mathbb{P}\mathcal{A}(\mathcal{W})^*$.

4.4 Double-Translation Hypersurfaces

By definition, a germ S of smooth hypersurface at $(\mathbb{C}^{n+1}, 0)$ is a **translation hypersurface** if it is non-degenerate and admits a regular parametrization of the form

$$\Phi \;:\; (\mathbb{C}^n, 0) \;\longrightarrow\; (\mathbb{C}^{n+1}, 0)$$

$$(x_1, \ldots, x_n) \longmapsto \phi_1(x_1) + \cdots + \phi_n(x_n)$$

---

[4]Recall that the canonical map of the projective curve C is defined as

$$\kappa_C : C \dashrightarrow \mathbb{P}H^0(C, \omega_C)^* \simeq \mathbb{P}^{g_a(C)-1}$$

$$x \longmapsto \mathbb{P}(\ker\{\omega \mapsto \omega(x)\}).$$

where $\phi_1, \ldots, \phi_{n+1} : (\mathbb{C}, 0) \to (\mathbb{C}^n, 0)$ are germs of holomorphic maps. Notice that such a parametrization Φ induces naturally a n-web $\mathcal{W}_\Phi = \Phi_* \mathcal{W}(x_1, \ldots, x_n)$ on S.

To understand the logic behind the terminology, notice that a surface S in $\mathbb{C}^3$ is a translation surface if it can be generated by translating a curve along another one (see Fig. 4.4).

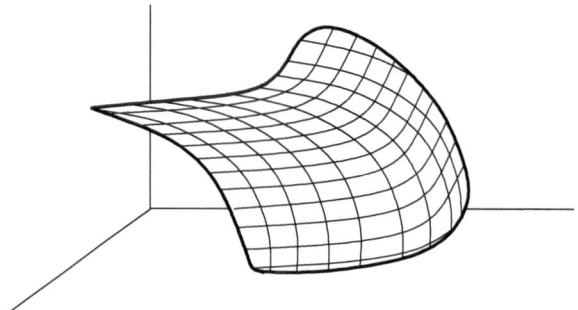

Fig. 4.4 A translation surface in $\mathbb{C}^3$

By definition, another parametrization

$$\Psi : (\mathbb{C}^n, 0) \longrightarrow (\mathbb{C}^{n+1}, 0)$$
$$(y_1, \ldots, y_n) \longmapsto \psi_1(y_1) + \cdots + \psi_n(y_n)$$

of S as a translation hypersurface is **distinct** from Φ if the superposition of the corresponding n-webs $\mathcal{W}_\Psi$ and $\mathcal{W}_\Phi$ is a quasi-smooth $2n$-web on S.

A hypersurface S is a **double-translation hypersurface** if it admits two distinct parametrizations in the above sense. The quasi-smooth $2n$-web $\mathcal{W}_\Phi \boxtimes \mathcal{W}_\Psi$ will be denoted by $\mathcal{W}_S$. Notice that there is a certain abuse of notation here since, a priori, a double-translation hypersurface may admit more than two distinct parametrizations as a translation hypersurface. Implicit in the notation $\mathcal{W}_S = \mathcal{W}_\Phi \boxtimes \mathcal{W}_\Psi$ is the fact that the two distinct parametrizations of translation type are fixed in the definition of a double-translation hypersurface.

Example 4.4.1. The surface S in $\mathbb{C}^3$ cut out by $4x + z^5 - 5zy^2 = 0$ is an example of an algebraic double-translation surface.

One verifies that

$$\Phi : (x_1, x_2) \longmapsto \left(x_1^{-5} + x_2^{-5}, x_1^{-2} + x_2^{-2}, x_1^{-1} + x_2^{-1} \right).$$

and

$$\Psi : (y_1, y_2) \longmapsto \left(-y_1^{-5} - y_2^{-5}, -y_1^{-2} - y_2^{-2}, -y_1^{-1} - y_2^{-1} \right)$$

are two parametrizations of S as a translation surface.

One verifies that, up to permutation of the variables, one has $\Phi(x_1, x_2) = \Psi(y_1, y_2)$ if and only if $(x_1, x_2) = (-y_1, -y_2)$ or $y_1 = x_1 x_2 \zeta_+$ and $y_2 = x_1 x_2 \zeta_-$ where $\zeta_\pm = \zeta_\pm(x_1, x_2)$ stand for the two complex roots of the polynomial $(x_1^2 + x_1 x_2 + x_2^2)\zeta^2 + (x_1 + x_2)\zeta + 1 = 0$, that is

$$\zeta_\pm(x_1, x_2) = -\frac{1}{2} \frac{x_1 + x_2 \pm \sqrt{-(3x_1^2 + 2x_1 x_2 + 3x_2^2)}}{x_1^2 + x_1 x_2 + x_2^2}.$$

If $p, q \in \mathbb{C}^2$ are points satisfying $\Phi(p) = \Psi(q)$ but no any linear relation, then $\Phi : (\mathbb{C}^2, p) \to S$ and $\Psi : (\mathbb{C}^2, q) \to S$ are two distinct parametrizations of S at $\Phi(p) = \Psi(q)$. Hence S is a double-translation hypersurface with

$$\mathcal{W}_S = \mathcal{W}\left(x_1, x_2, x_1 x_2 \zeta_-, x_1 x_2 \zeta_+\right).$$

4.4.1 Examples

The preceding example is a particular instance of a general construction presented below.

Let $C \subset \mathbb{P}^n$ be a reduced non-degenerate projective curve of degree $2n$. Assume that $h^0(\omega_C) \geq n + 1$. If C is $\mathcal{W}$-generic, then $h^0(\omega_C) = n + 1$, but otherwise $h^0(\omega_C)$ can be larger.

Let H_0 be a hyperplane intersecting C transversely at $2n$ points. As usual, consider the holomorphic germs $p_1, \ldots, p_{2n} : (\check{\mathbb{P}}^n, H_0) \to C$ satisfying $C \cdot H = \sum_{i=1}^{2n} p_i(H)$ for every $H \in (\check{\mathbb{P}}^n, H_0)$. Since C is non-degenerate, it is harmless to assume that the p_i's have been indexed in such a way that $p_1(H), \ldots, p_n(H)$ generate H for every $H \in (\check{\mathbb{P}}^n, H_0)$.

If $\omega^1, \ldots \omega^{n+1}$ are linearly independent abelian differentials on C, then for each $i \in \underline{n}$ consider the maps $\phi_i, \psi_i : (\check{\mathbb{P}}^n, H_0) \to \mathbb{C}^{n+1}$ defined as

$$\phi_i(H) = \left(\int_{p_i(H_0)}^{p_i(H)} \omega^1, \ldots, \int_{p_i(H_0)}^{p_i(H)} \omega^{n+1} \right)$$

and

$$\psi_i(H) = -\left(\int_{p_{i+n}(H_0)}^{p_{i+n}(H)} \omega^1, \ldots, \int_{p_{i+n}(H_0)}^{p_{i+n}(H)} \omega^{n+1} \right).$$

Notice that their sums for i ranging from 1 to n, that is $\Phi = \sum_{i=1}^n \phi_i$ and $\Psi = \sum_{i=1}^n \psi_i$, satisfy $\Phi(H) = \Psi(H)$ for every $H \in (\check{\mathbb{P}}^n, H_0)$ according to Abel's addition Theorem. Furthermore, they parametrize a non-degenerate hypersurface $S_C \subset (\mathbb{C}^n, 0)$, and $\mathcal{W}_{S_C} = \mathcal{W}_\Phi \boxtimes \mathcal{W}_\Psi$ is a $2n$-web equivalent to the algebraic web $\mathcal{W}_C$ associated with C. From all that, it is clear that S_C is a double-translation hypersurface. By definition, it is a **double-translation hypersurface associated with C at H_0**. The use of the indefinite article "*a*" instead of the definite one "*the*" is due to the lack of uniqueness of the hypersurface for C and H_0 fixed. It will depend on the choice of the 1-forms and of the points p_i in general. When C is irreducible, a monodromy argument (see [67, p. 39] for instance) shows that it is possible to replace the "a" by a "the."

4.4.2 Abel–Jacobi Map

The double-translation hypersurface associated with an irreducible non-degenerate projective curve $C \subset \mathbb{P}^n$ of degree $2n$ and arithmetic genus $n + 1$ admits a more intrinsic description which has the advantage of being global. It is defined in terms of the Abel–Jacobi map of C. Although much of the discussion can be carried out in greater generality, this will not be done here. The interested reader can consult [85].

If C_{sm} stands for the smooth part of C, then there is a linear map from $H_1(C_{sm}, \mathbb{Z})$ to $H^0(C, \omega_C)^*$ defined as

$$\gamma \longmapsto \left(\omega \mapsto \int_\gamma \omega \right).$$

It can be shown that its image Γ is a discrete subgroup of $H^0(C, \omega_C)^*$. Consequently, the quotient of $H^0(C, \omega_C)^*$ by Γ is a smooth complex variety $J(C)$: the **Jacobian** of C. When C is smooth then $J(C)$ is indeed projective, as was shown by Riemann (cf. [63, Chapter 3] for instance).

Once a point $p \in C_{sm}$ and a positive integer k are fixed, one can consider the map

$$AJ_C^k : (C_{sm})^k \longrightarrow J(C)$$

$$(x_1, \dots, x_k) \longmapsto \left(\omega \mapsto \sum_{i=1}^k \int_p^{x_i} \omega \right),$$

where the integrations are performed along paths included in C_{sm}. Notice that AJ_C^k is well defined since all possible ambiguities disappear after taking the quotient by Γ. By definition, AJ_C^k is the kth **Abel–Jacobi map** of C.

If C is a smooth projective curve of genus $n + 1$, then a classical theorem of Riemann (see [7, Chap. I.§5]) asserts that the image of the nth Abel–Jacobi map is a translate of the theta divisor Θ of $J(C)$, which is, by definition, the reduced divisor defined as the zero locus of *Riemann's theta function* θ of the jacobian $J(C)$.[5]

---

[5]The **Riemann's theta function** θ_A of a polarized abelian variety $A = \mathbb{C}^g / \Delta$ with $\Delta = (I_g, Z)$ (where $Z \in M_g(\mathbb{C})$ is such that $Z = {}^t Z$ and $\operatorname{Im} Z > 0$) is defined by $\theta_A(z) = \sum_{m \in \mathbb{Z}^g} \exp \left(i\pi \langle m, Zm \rangle + 2i\pi \langle m, z \rangle \right)$ for all $z \in \mathbb{C}^g$ (see [7, Chap. I]).

Proposition 4.4.2. *If $C \subset \mathbb{P}^n$ is a smooth non-degenerate curve of degree $2n$ and genus $n + 1$, then the double-translation hypersurface S_C associated with C is nothing but the lift to $\mathbb{C}^{n+1}$ of (a translate of) the theta divisor $\Theta \subset J(C)$.*

Proof. From Riemann's result referred to above it suffices to show that S_C can be identified with the image of the nth Abel–Jacobi map.

Since C is non-degenerate, $P = (p_1, \ldots, p_n) : (\check{\mathbb{P}}^n, H_0) \to C^n$ is a germ of biholomorphism. Moreover, Φ is, up to a suitable choice of affine coordinates, equal to $AJ_C^n \circ P$ (or rather, to its natural lift to $H^0(C, \Omega_C^1)^* \simeq \mathbb{C}^{n+1}$). The proposition follows.   □

When C is smooth, it is known that the lift of Θ in $\mathbb{C}^{n+1}$ is a transcendental hypersurface. When C is singular, the hypersurface S_C is not necessarily transcendent. Loosely speaking, the more C is singular, the less S_C is transcendent. For instance, we let the reader verify that the rational double-translation surface of Example 4.4.1 is nothing but the surface S_C associated with the plane singular rational quartic parametrized by $\mathbb{P}^1 \ni [s : t] \mapsto [s^4 : st^3 : t^4] \in \mathbb{P}^2$ (see also Example 3.2.4 in Chap. 3).

4.4.3 Classification

Theorem 4.4.3. *Let $S \subset \mathbb{C}^{n+1}$ be a non-degenerate double translation hypersurface such that $\mathcal{W}_S$ is smooth. Then S is the double-translation hypersurface associated with a $\mathcal{W}$-generic projective curve in $\mathbb{P}^n$ of degree $2n$ and arithmetic genus $n + 1$.*

Proof. Let Φ and Ψ be two distinct translation-type parametrizations of a double-translation hypersurface $S \subset \mathbb{C}^{n+1}$. The coordinates $x_1, \ldots, x_n, y_1, \ldots, y_n$ in which Φ and Ψ are expressed can be seen as holomorphic functions on S defining the web $\mathcal{W}_S$.

Hence the identity

$$\sum_{i=1}^{n} \Phi(x_i(p)) + \sum_{i=1}^{n} \Psi(y_i(p)) = 0 \in \mathbb{C}^{n+1}$$

holds at any point $p \in S$. Since S is non-degenerate this equation provides $n + 1 = \pi(n, 2n)$ linearly independent abelian relations for $\mathcal{W}_S$. Theorem 4.3.4 implies the result.   □

When $\mathcal{W}_S$ is only assumed to be quasi-smooth, one has the following algebraization result which can be traced back to Wirtinger [134].

Theorem 4.4.4. *Let $S \subset (\mathbb{C}^{n+1}, 0)$ be a double-translation hypersurface. Assume that its distinguished parametrizations $\Phi = (\phi_i)_{i=1}^{n}$ and $\Psi = (\psi_i)_{i=1}^{n}$ are such that*

($\star$) *none of the vectors $\frac{d^2\phi_i}{dx_i^2}(0), \frac{d^2\psi_i}{dy_i^2}(0)$ is tangent to S at the origin.*

Then S is the double-translation hypersurface associated with a non-degenerate curve $C \subset \mathbb{P}^n$ of degree $2n$ and such that $h^0(\omega_C) \geq n + 1$.

For a proof of the above result, the reader is redirected to [85] from where the formulation above has been borrowed. There he will also find the following application to the Schottky Problem: characterize the Jacobian of curves among the principally polarized abelian varieties.

Theorem 4.4.5. *Let (A, Θ) be a principally polarized abelian variety of dimension n. Suppose that there exists a point $p \in \Theta$ such that the germ of Θ at p is a double-translation hypersurface satisfying ($\star$). Then (A, Θ) is the canonically polarized Jacobian of a smooth non-hyperelliptic curve of genus n.*

For more information about the Schottky Problem the reader is urged to consult [12, 88, Appendix, Lecture IV] and the references therein.

Chapter 5
Algebraization of Maximal Rank Webs

This chapter is devoted to the following result.

Theorem. *Let $n \geq 3$ and $k \geq 2n$ be integers. If $\mathcal{W}$ is a smooth k-web of maximal rank on $(\mathbb{C}^n, 0)$, then $\mathcal{W}$ is algebraizable.*

Its proof crowns the efforts spreaded over at least three generations of mathematicians. For $n = 3$, the theorem is due to Bol and is among the deepest results obtained by Blaschke's school. For $n > 3$, Chern and Griffiths provided a "proof" in [34] which later on revealed itself to be incomplete [36]. The definitive version stated above is due to Trépreau [130]. He did not just follow Chern–Griffiths's general strategy, but he also simplified it, to prove the general case.

While many of the concepts and ideas used in the proof—Poincaré's map and canonical maps—have already been introduced in Chap. 4, it will be essential to consider also another map, here called Poincaré–Blaschke's map, canonically attached to webs of maximal rank. This map was originally introduced by Blaschke in [16] to prove that planar 5-webs of maximal rank are algebraizable. Using Blaschke's ideas introduced in this paper, Bol succeeded to establish the algebraization of smooth k-webs of maximal rank on $(\mathbb{C}^3, 0)$, for $k \geq 6$. Ironically, not much later [21], he came up with $6 = \pi(2, 5)$ linearly independent abelian relations for his non-algebraizable 5-web $\mathcal{B}_5$, thus showing that the source of his inspiration [16] was irremediably flawed. Although wrong, Blaschke's paper contained not just the germ of Bol's algebraization result for webs on $(\mathbb{C}^3, 0)$, but also the germ of Chern–Griffiths's strategy.

The main novelty in Trépreau's approach is based on ingenious and involved, albeit elementary, computations. It is still considerably more technical than the other results previously presented in this book.

© Springer International Publishing Switzerland 2015
J.V. Pereira, L. Pirio, *An Invitation to Web Geometry*, IMPA Monographs 2,
DOI 10.1007/978-3-319-14562-4_5

5.1   Trépreau's Theorem

In this section, a stronger version of the theorem stated in the introduction of this
chapter is formulated. Then the heuristic behind its proof is explained.

Arguably, one could complain about the unfairness of the title of this section.
It would perhaps be more righteous to call it Bol–Trépreau's Theorem or even
Blaschke–Bol–Chern–Grifitths–Trépreau's Theorem. To justify the choice made
above, one could invoke the right to typographical beauty and/or the usual mathematical practice.

5.1.1   Statement

Theorem 5.1.1. *Let* $\mathcal{W}$ *be a smooth k-web on* $(\mathbb{C}^n, 0)$. *If $n \geq 3$, $k \geq 2n$ and*

$$\dim \frac{\mathcal{A}(\mathcal{W})}{F^2\mathcal{A}(\mathcal{W})} = 2k - 3n + 1,$$

then $\mathcal{W}$ *is algebraizable.*

To explain the heuristic behind the proof of Theorem 5.1.1 some aspects of the
geometry of Castelnuovo curves will be recalled in Sect. 5.1.2. Then, in Sect. 5.1.3,
we will discuss how one could infer corresponding geometrical properties for webs
of maximal rank using their spaces of abelian relations.

5.1.2   The Geometry of Castelnuovo Curves

Let $C \subset \mathbb{P}^n$ be a Castelnuovo curve of degree $k \geq 2n$. To avoid technicalities
assume that C is smooth. By definition $K_C = \omega_C$ and $h^0(K_C) = \pi(n, k) = \pi > 0$.

Recall from Chap. 4 that the canonical map of C is

$$\varphi = \varphi_{|K_C|} : C \longrightarrow \mathbb{P}H^0(C, K_C)^* = \mathbb{P}^{\pi-1} \tag{5.1}$$

$$x \longmapsto [\eta \mapsto \eta(x)] .$$

Since C is smooth of genus strictly greater than one, the canonical linear system
$|K_C|$ has no base point (*cf.* [70, IV.5]). Moreover Castelnuovo curves are not
hyperelliptic and thus (see [70, IV.5]) $\varphi = \varphi_{|K_C|}$ is a birational morphism from
C onto its **canonical model**

$$C_{\text{can}} = \varphi(C) \subset \mathbb{P}^{\pi-1}.$$

Recall from Sect. 3.3 that the linear system $|I_C(2)|$ of quadrics containing C cuts out a non-degenerate surface $S \subset \mathbb{P}^n$ of minimal degree $n - 1$. Thus, according to Theorem 2.4.13, S is a rational normal scroll or the Veronese surface $v_2(\mathbb{P}^2) \subset \mathbb{P}^5$ (notice that when $n = 3$, S is a quadric surface hence nothing else than one of the two scrolls $S_{1,1}$ or $S_{0,2}$).

Still aiming at simplicity, assume that S is smooth. Because S is rational, it does not have non-zero holomorphic differentials. More precisely, $h^0(S, \Omega_S^2) = h^0(S, K_S) = 0$.

Consequently, one has

$$h^1(S, K_S) = h^1(S, \mathcal{O}_S) \qquad \text{by Serre duality [70, III.7]}$$

$$= h^0(S, \Omega_S^1) = 0 \qquad \text{by Hodge symmetry.}$$

Applying these vanishing results to the long cohomology sequence associated with the exact sequence

$$0 \to K_S \to K_S \otimes \mathcal{O}_S(C) \to K_C \to 0,$$

one deduces an isomorphism

$$H^0\big(S, K_S \otimes \mathcal{O}_S(C)\big) \simeq H^0\big(C, K_C\big).$$

Thus the canonical map (5.1) extends to S: there is a rational map $\Phi : S \dashrightarrow \mathbb{P}^{\pi-1}$ fitting into the commutative diagram below.

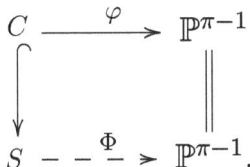

Moreover, as it is explained in [67], X_C, the image of Φ, is a non-degenerate algebraic surface in $\mathbb{P}^{\pi-1}$.

If H is a generic hyperplane in $\mathbb{P}^n$, then the hyperplane section $C_H = S \cap H$ is irreducible, non-degenerate in $H \simeq \mathbb{P}^{n-1}$, and of degree $\deg(C_H) = \deg(S) = n - 1$. Hence C_H is a curve of minimal degree in H, that is a rational normal curve of degree $n - 1$.

Now let $s = (s_1, \dots, s_n) \in S^n$ be a generic n-uplet of points on S. The points s_i span a hyperplane H_s and the general hyperplane is obtained in this way. Thus $C_{H_s} = S \cap H_s$ is a rational normal curve of degree $n - 1$ contained in S that passes through the points $s_1, \dots, s_n$. Therefore, through n general points of S passes a rational normal curve of degree $n - 1$ included in S. Surfaces having this property are said to be n-**covered by rational normal curves of degree** $n - 1$.

It can be proved that the image under Φ of C_H is a rational normal curve $\mathscr{C}_H$ in a projective subspace of $\mathbb{P}H^0(C, K_C)^* \simeq \mathbb{P}^{\pi-1}$ of dimension $k - n - 1$. Consequently, the surface $X_C = \mathrm{Im}(\Phi) \subset \mathbb{P}^{\pi-1}$ is n-covered by rational normal curves of degree $k - n - 1$.

The preceding facts will now be interpreted in terms of the web $\mathcal{W}_C$ dual to the curve C. As usual, let $H_0 \subset \mathbb{P}^n$ be a hyperplane transverse to C, and let $p_1, \ldots, p_k : (\check{\mathbb{P}}^n, H_0) \to C$ be the usual holomorphic maps describing the intersection of $H \in (\check{\mathbb{P}}^n, H_0)$ with C.

For $i \in \underline{k}$, set

$$\varphi_i = \varphi_{|K_C|} \circ p_i : (\check{\mathbb{P}}^n, H_0) \to \mathbb{P}H^0(C, K_C)^*.$$

If $H \in (\check{\mathbb{P}}^n, H_0)$, then the points $p_1(H), \ldots, p_k(H)$ span the hyperplane $H \subset \mathbb{P}^n$. Thus they belong to C, and hence to S. Therefore each $p_i(H)$ belongs to the rational normal curve $C_H = C \cap H \subset S$. Consequently the points $\varphi_i(H)$, $i \in \underline{k}$, belong to the rational normal curve $\mathscr{C}_H \subset \mathbb{P}H^0(C, K_C)^*$. Moreover, the curves $\mathscr{C}_H$ for $H \in (\check{\mathbb{P}}^n, H_0)$ fill out a germ of surface along $\mathscr{C}_{H_0}$. All this is illustrated in Fig. 5.1.

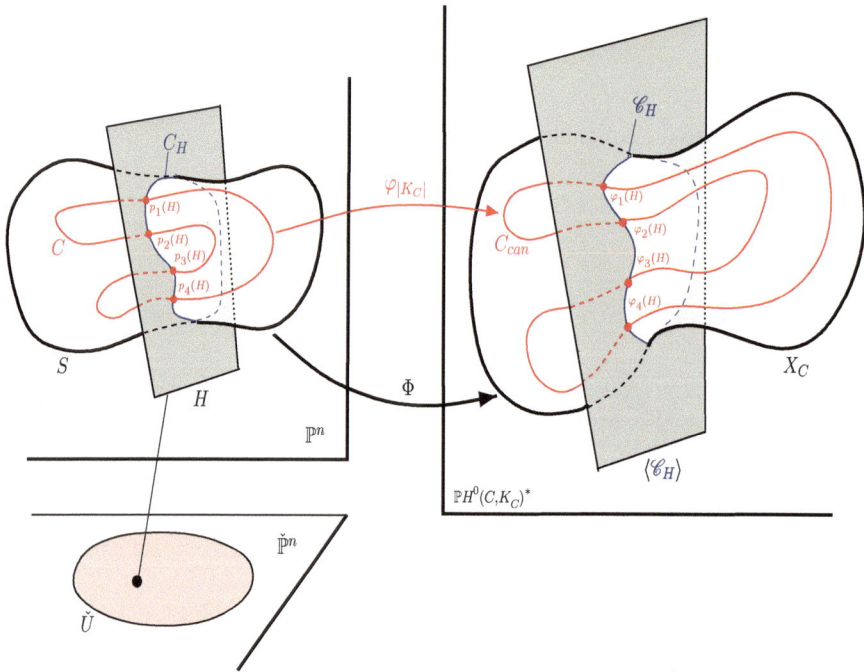

Fig. 5.1 The curve C_H and its image $\mathscr{C}_H$ under Φ

5.1.3 On the Geometry of Maximal Rank Webs

We now explain how the constructions presented in the preceding section extend to webs carrying sufficiently many abelian relations. Since the case $k = 2n$ has already been treated in Chap. 4, it will be assumed that $k > 2n$ in what follows.

Let $\mathcal{W}$ be a smooth k-web on $(\mathbb{C}^n, 0)$. To simplify the discussion, it will be assumed that $\mathcal{W}$ has maximal rank, even if the heuristic described below will be implemented under the weaker hypothesis of Theorem 5.1.1.

Using the hypothesis on the space of abelian relations of $\mathcal{W}$, it can be proved that the images $\kappa_{\mathcal{W},1}(x), \ldots, \kappa_{\mathcal{W},k}(x)$ of every $x \in (\mathbb{C}^n, 0)$ by the canonical maps of $\mathcal{W}$ generate a projective subspace $\mathbb{P}^{k-n-1}(x) \subset \mathbb{P}\mathcal{A}(\mathcal{W})^*$ of dimension $k - n - 1$, and lie on a unique rational normal curve $\mathscr{C}(x) \subset \mathbb{P}^{k-n-1}(x)$ of degree $k - n - 1$.

The most delicate point, and the main novelty, in Trépreau's argument, is his proof that the family of rational normal curves $\mathscr{C}_{\mathcal{W}} = \{\mathscr{C}(x)\}_{x \in (\mathbb{C}^n, 0)}$ fills out a germ of non-degenerate, smooth surface $X_{\mathcal{W}} \subset \mathbb{P}\mathcal{A}(\mathcal{W})^*$ along $\mathscr{C}(0)$ (Fig. 5.2).

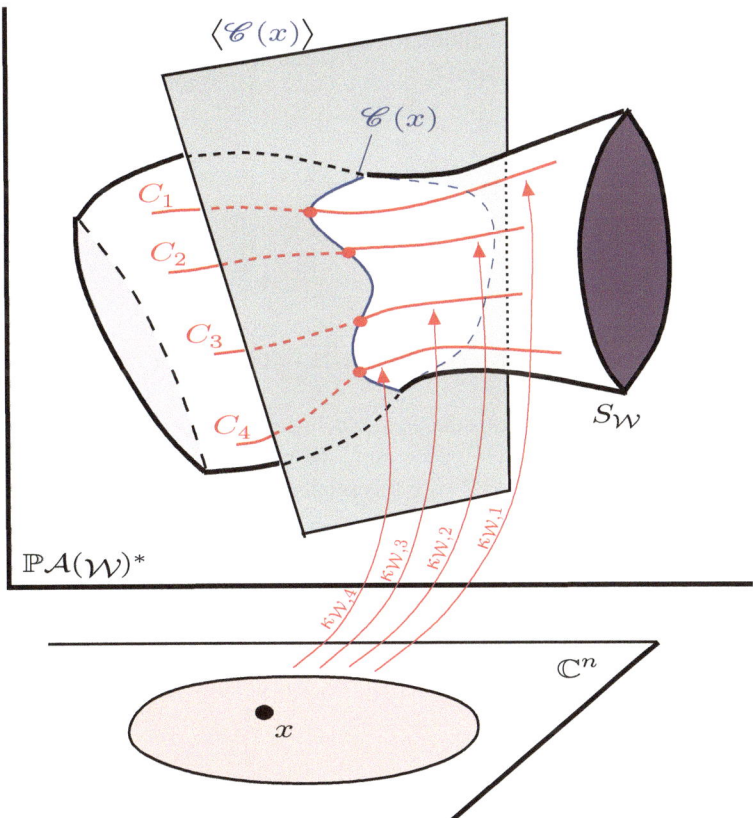

Fig. 5.2 Geometry behind the proof of Trépreau's Theorem

It is then comparably simpler to show that any two curves $\mathscr{C}(x), \mathscr{C}(y)$ intersect in exactly $n-1$ points when $x, y \in (\mathbb{C}^n, 0)$ are distinct. Therefore $\mathscr{C}(0)^2 = n - 1 > 0$ and well-known results about germs of surfaces containing curves of positive self-intersection by Andreotti (in the analytic category) and Hartshorne (in the algebraic category) imply that $X_{\mathcal{W}}$ is contained in a projective surface $S_{\mathcal{W}}$. It can be proved that, as the image under Φ of the surface of minimal degree in $\mathbb{P}^n$ containing a Castelnuovo curve C of degree k (see the previous section), the surface $S_{\mathcal{W}}$ is a rational surface n-covered by rational normal curves of degree $k - n - 1$.

At this point, one can apply an argument by Chern and Griffiths to linearize $\mathcal{W}$. Since $S_{\mathcal{W}}$ is rational, every curve in the family $\mathscr{C}_{\mathcal{W}} = \{\mathscr{C}(x)\}_{x \in (\mathbb{C}^n, 0)}$ belongs to one and only one linear system

$$|\mathscr{C}| = \mathbb{P} H^0\big(S_{\mathcal{W}}, \mathcal{O}_{S_{\mathcal{W}}}(\mathscr{C}(0))\big)$$

which turns out to have dimension n.

One then defines a map $\mathfrak{C} : (\mathbb{C}^n, 0) \to \mathbb{P}^n$ sending $x \in (\mathbb{C}^n, 0)$ to the point in $|\mathscr{C}|$ corresponding to the rational normal curve $\mathscr{C}(x)$. It is possible to prove that this map is a biholomorphism which linearizes $\mathcal{W}$. Finally, the converse to Abel's Theorem presented in Chap. 4 allows to conclude that $\mathfrak{C}_* \mathcal{W}$ is algebraic. This shows in particular that $\mathcal{W}$ is algebraizable.

5.2 Maps Naturally Attached to $\mathcal{W}$

Until the end of this chapter $\mathcal{W} = \mathcal{F}_1 \boxtimes \cdots \boxtimes \mathcal{F}_k$ is a smooth k-web on $(\mathbb{C}^n, 0)$ satisfying

$$\dim \mathcal{A}(\mathcal{W})/F^2 \mathcal{A}(\mathcal{W}) = 2k - 3n + 1 = \pi .$$

According to this hypothesis there exist π linearly independent abelian relations with classes in $\mathcal{A}(\mathcal{W})/F^2 \mathcal{A}(\mathcal{W})$ generating the whole space. Fix, once and for all, π abelian relations $\eta^{(1)}, \ldots, \eta^{(\pi)}$ with this property and let $\mathcal{A}_2(\mathcal{W}) \subset \mathcal{A}(\mathcal{W})$ be the vector space generated by them.

Notice that the inclusion $\mathcal{A}_2(\mathcal{W}) \subset \mathcal{A}(\mathcal{W})$ induces a linear projection $\mathbb{P} \mathcal{A}(\mathcal{W})^* \dashrightarrow \mathbb{P} \mathcal{A}_2(\mathcal{W})^*$. Notice also that the intersection of the filtration $F^\bullet \mathcal{A}(\mathcal{W})$ of $\mathcal{A}(\mathcal{W})$ with $\mathcal{A}_2(\mathcal{W})$ induces the filtration

$$F^\bullet \mathcal{A}_2(\mathcal{W}) = \mathcal{A}_2(\mathcal{W}) \cap F^\bullet \mathcal{A}(\mathcal{W}).$$

From the choice of $\mathcal{A}_2(\mathcal{W})$ it is clear that

$$\dim F^1 \mathcal{A}_2(\mathcal{W}) = k - 2n + 1 \quad \text{and} \quad \dim F^j \mathcal{A}_2(\mathcal{W}) = 0 \text{ for } j > 1. \tag{5.2}$$

Following the convention of Sect. 4.3.1, we will denote by $F_x^\bullet \mathcal{A}_2(\mathcal{W})$ the corresponding filtration of $\mathcal{A}_2(\mathcal{W})$ induced by the germs of $\eta^{(1)}, \ldots, \eta^{(\pi)}$ at $x \in (\mathbb{C}^n, 0)$.

5.2.1   Canonical Maps

Let $i \in \underline{k}$ be fixed. If $\kappa_i = \kappa_{\mathcal{W},i} : (\mathbb{C}^n, 0) \to \mathbb{P}\mathcal{A}(\mathcal{W})^*$ is the ith canonical map of $\mathcal{W}$, see Sect. 4.3.5, then its image does not intersect the center of the natural projection $\mathbb{P}\mathcal{A}(\mathcal{W})^* \dashrightarrow \mathbb{P}\mathcal{A}_2(\mathcal{W})^*$. Indeed, as argued in the proof of Lemma 4.3.1, the equality $\dim \mathcal{A}_2(\mathcal{W})/F^1\mathcal{A}_2(\mathcal{W}) = k - n$ implies the existence of an abelian relation in $\mathcal{A}_2(\mathcal{W})$ whose ith component does not vanish at 0. Thus composing κ_i with the linear projection $\mathbb{P}\mathcal{A}(\mathcal{W})^* \dashrightarrow \mathbb{P}\mathcal{A}_2(\mathcal{W})^*$ defines a morphism from $(\mathbb{C}^n, 0)$ to $\mathbb{P}\mathcal{A}_2(\mathcal{W})^*$. To keep the notation simple, it will still be denoted by κ_i and will again be called the ith **canonical map of** $\mathcal{W}$.

More explicitly, κ_i now stands for the map

$$(\mathbb{C}^n, 0) \longrightarrow \mathbb{P}\mathcal{A}_2(\mathcal{W})^*$$
$$x \longmapsto \big[ev_i(x) : \mathcal{A}_2(\mathcal{W}) \to \mathbb{C}\big].$$

The image of the ith canonical map of $\mathcal{W}$ will be denoted by C_i and will be called the ith **canonical curve of** $\mathcal{W}$. One has:

$$C_i = \mathrm{Im}(\kappa_i) \subset \mathbb{P}\mathcal{A}_2(\mathcal{W})^*.$$

In order to express intrinsically the differential of κ_i at a point $x \in (\mathbb{C}^n, 0)$, observe that the tangent space of $\mathbb{P}\mathcal{A}_2(\mathcal{W})^*$ at the point $\kappa_i(x) = [ev_i(x)]$ is naturally isomorphic to the quotient of $\mathcal{A}_2(\mathcal{W})^*$ by the one-dimensional subspace $\mathbb{C}ev_i(x)$. This quotient in turn is isomorphic to the dual of $\ker ev_i(x)$. Thus the differential of κ_i at x can be written as follows:

$$d\kappa_i(x) : T_x(\mathbb{C}^n, 0) \longrightarrow \ker\big(ev_i(x)^*\big)$$
$$v \longmapsto \left\{ v \mapsto \lim_{t \to 0} \frac{ev_i(x + tv) - ev_i(x)}{t} \right\}.$$

Therefore, inasmuch $\kappa_i(x)$ can be identified with the space of abelian relations in $\mathcal{A}_2(\mathcal{W})$ whose ith component vanishes at x, the image of its differential at x can be identified with the space of abelian relations in $\mathcal{A}_2(\mathcal{W})$ with ith component vanishing at x with multiplicity two. In other words, if $ev_i^{(1)}(x) : \mathcal{A}_2(\mathcal{W}) \to \mathbb{C}^2$ is the evaluation morphism of order one (as defined in the proof of Proposition 4.3.5) and if $V_i(x) \subset \mathcal{A}_2(\mathcal{W})$ stands for its kernel, then the image of $d\kappa_i(x)$ is the quotient of $\ker(\mathcal{A}_2(\mathcal{W})^* \to V_i(x)^*)$ by $\mathbb{C}ev_i(x)$.

Lemma 5.2.1. *For every* $i \in \underline{k}$, *the canonical map* $\kappa_i : (\mathbb{C}^n, 0) \to C_i$ *is a submersion.*

Proof. According to the discussion preceding the statement, it suffices to prove that the inclusion $V_i(0) \subset \mathrm{Ker}(ev_i(0))$ is proper. Assuming that it is not the case, one would have in particular

$$F^1 \mathcal{A}_2(\mathcal{W}) = F^1 \mathcal{A}_2(\mathcal{W}) \cap V_i(0).$$

But

$$\dim F^1 \mathcal{A}_2(\mathcal{W}) \cap V_i(0) \leq (k-1) - \ell^2 \big(\boxtimes_{j \in \underline{k} \backslash \{i\}} \mathcal{F}_j \big) + \dim F^2 \mathcal{A}_2(\mathcal{W}).$$

Since $F^2 \mathcal{A}_2(\mathcal{W}) = 0$ and because $\ell^2 \big(\boxtimes_{j \in \underline{k} \backslash \{i\}} \mathcal{F}_j \big) \geq 2(n-1) + 1$ according to Proposition 2.3.1, this would imply $\dim F^1 \mathcal{A}_2(\mathcal{W}) \leq k - 2n$, which would contradict (5.2). $\square$

Actually, the preceding proof proves more than the statement of Lemma 5.2.1, cf. the last part of Lemma 5.2.3 below.

5.2.2 Poincaré's Map

Consider the natural analogue of Poincaré's map

$$P_{\mathcal{W}} : (\mathbb{C}^n, 0) \longrightarrow \mathrm{Grass}(\mathcal{A}_2(\mathcal{W}), k - 2n + 1)$$

$$x \longmapsto F_x^1 \mathcal{A}_2(\mathcal{W}).$$

As in the case of canonical maps, it seems unjustifiable to change the terminology. Therefore the map $P_{\mathcal{W}}$ above will also be called the Poincaré's map of $\mathcal{W}$.

For each $x \in (\mathbb{C}^n, 0)$, the projective subspace of dimension $k - n - 1$ in $\mathbb{P}\mathcal{A}_2(\mathcal{W})^*$ determined by $P_{\mathcal{W}}(x)$ through projective duality will be denoted by $\mathbb{P}^{k-n-1}(x)$.

5.2.3 Some Properties of $P_{\mathcal{W}}$

From the very definition of $F_x^1 \mathcal{A}_2(\mathcal{W})$ it follows that

$$F_x^1 \mathcal{A}_2(\mathcal{W}) = \bigcap_{i \in \underline{k}} \ker \big(ev_i(x) \big)$$

where $ev_i(x)$ is considered here as a linear form on $\mathcal{A}_2(\mathcal{W})$.

This remark about subspaces of $\mathcal{A}_2(\mathcal{W})$ translates into the following relation between the canonical maps and Poincaré's map: $\mathbb{P}^{k-n-1}(x)$ is the smallest projective space among the ones containing the set $\{\kappa_i(x)\}_{i \in \underline{k}}$, i.e. one has

$$\mathbb{P}^{k-n-1}(x) = \langle \kappa_1(x), \ldots, \kappa_k(x) \rangle \subset \mathbb{P}^{\pi-1}.$$

The lemma below shows that it is possible to replace $\underline{k}$, in the statement above, by any subset B with at least $k - n$ elements.

Lemma 5.2.2. *For every $x \in (\mathbb{C}^n, 0)$ and every subset $B \subset \underline{k}$ of cardinality $k - n$, $\mathbb{P}^{k-n-1}(x)$ is the smallest linear subspace of $\mathbb{P}\mathcal{A}_2(\mathcal{W})^*$ containing $\{\kappa_i(x)\}_{i \in B}$, i.e.*

$$\mathbb{P}^{k-n-1}(x) = \langle \kappa_i(x) \,|\, i \in B \rangle \subset \mathbb{P}^{\pi-1}.$$

Proof. Notice that the smallest linear subspace of $\mathbb{P}\mathcal{A}_2(\mathcal{W})^*$ containing $\{\kappa_i(x)\}_{i \in B}$ is the dual of the intersection

$$I_x = \bigcap_{i \in B} \left[\ker ev_i(x) \right] \subset \mathbb{P}\mathcal{A}_2(\mathcal{W}).$$

If a non-trivial abelian relation, or rather its projectivization, is in I_x, then it has at most $k - (k - n) = n$ components not vanishing at x. But the constant terms of these components are linearly independent because $\mathcal{W}$ is a smooth web on $(\mathbb{C}^n, 0)$. Thus

$$I_x = \bigcap_{i \in B} \left[\ker ev_i(x) \right] = \left[F_x^1 \mathcal{A}_2(\mathcal{W}) \right] \subset \mathbb{P}\mathcal{A}_2(\mathcal{W}). \qquad \square$$

Lemma 5.2.3. *Let $x \in (\mathbb{C}^n, 0)$ and $B \subset \underline{k}$ be a subset of cardinality smaller than or equal to $k-2n+1$. Then the smallest projective subspace of $\mathbb{P}\mathcal{A}_2(\mathcal{W})^*$ containing*

$$\mathbb{P}^{k-n-1}(x) \cup \left(\cup_{i \in B} \, T_{\kappa_i(x)} C_i \right) \subset \mathbb{P}\mathcal{A}_2(\mathcal{W})^*$$

has codimension equal to $(k - 2n + 1) - \mathrm{card}(B)$. In particular, for every $i \in \underline{k}$, the canonical curve C_i intersects $\mathbb{P}^{k-n-1}(x)$ transversely at $\kappa_i(x)$.

Proof. The proof is a generalization of the one of Lemma 5.2.1.

Let $x \in (\mathbb{C}^n, 0)$ be fixed and, as in the discussion preceding Lemma 5.2.1, let $V_i = V_i(x)$ be the kernel of the evaluation morphism of order one $ev_i^{(1)}(x) : \mathcal{A}_2(\mathcal{W}) \to \mathbb{C}^2$.

Since $F_x^1 \mathcal{A}_2(\mathcal{W}) = \bigcap_{i \in \underline{k}} \ker(ev_i(x))$, to prove the lemma it suffices to show that the dimension, denoted by a, of

$$A = F_x^1 \mathcal{A}_2(\mathcal{W}) \cap \left(\cap_{i \in B} V_i \right)$$

is equal to $k - 2n + 1 - \mathrm{card}(B)$.

Since $\dim F_x^1 \mathcal{A}_2(\mathcal{W}) = k - 2n + 1$ and $\dim F_x^1 \mathcal{A}_2(\mathcal{W})/(F_x^1 \mathcal{A}_2(\mathcal{W}) \cap V_i(x)) \geq 1$ for every $i \in \underline{k}$, the number a is greater than or equal to $k - 2n + 1 - \mathrm{card}(B)$.

Notice that the elements of A are abelian relations with components vanishing at x, and with ith component having constant and linear terms at x equal to zero. Therefore,

$$a \leq (k - \mathrm{card}(B)) - \ell^2 \big(\boxtimes_{i \in \underline{k} \setminus B} \mathcal{F}_i \big) + \dim F_x^2 \mathcal{A}_2(\mathcal{W}) \,.$$

But $\dim F_x^2 \mathcal{A}_2(\mathcal{W}) = 0$ and Proposition 2.3.1 implies

$$\ell^2 \big(\boxtimes_{i \in \underline{k} \setminus B} \mathcal{F}_i \big) \geq 2(n - 1) + 1 \,.$$

Thus

$$a \leq \big(k - \mathrm{card}(B) \big) - 2(n - 1) - 1 = k - 2n + 1 - \mathrm{card}(B) \,,$$

as wanted. $\square$

Lemma 5.2.3 for subsets $B \subset \underline{k}$ of cardinality one is, together with Lemma 5.2.2, the main ingredient in the proof of the next proposition.

Proposition 5.2.4. *Poincaré's map $P_{\mathcal{W}}$ is an immersion.*

Proof. Let $\gamma : (\mathbb{C}, 0) \to (\mathbb{C}^n, 0)$ be a holomorphic immersion. Since $\mathcal{W}$ is smooth, γ is tangent to at most $n - 1$ of the foliations $\mathcal{F}_i$. Thus there exists a set $B \subset \underline{k}$ of cardinality $k - n \leq k - (n - 1)$, such that the composition $\kappa_i \circ \gamma$ is an immersion for every $i \in B$. Moreover, Lemma 5.2.2 implies that for every $t \in (\mathbb{C}, 0)$, the points $\{(\kappa_i \circ \gamma)(t)\}_{i \in B}$ generate the projective subspace $\mathbb{P}^{k-n-1}(\gamma(t)) \subset \mathbb{P}\mathcal{A}_2(\mathcal{W})^*$ determined by $P_{\mathcal{W}}(\gamma(t))$.

If one identifies the Plücker embeddings[1] of the grassmannians $\mathrm{Grass}(\mathcal{A}_2(\mathcal{W}), k - 2n + 1)$ and $\mathrm{Grass}(\mathcal{A}_2(\mathcal{W})^*, k - n)$, then one can write

$$(\widehat{P_{\mathcal{W}}} \circ \gamma)(t) = \bigwedge_{i \in B} (\widehat{\kappa_i} \circ \gamma)(t) \,,$$

where the hats indicate liftings to $\bigwedge^{k-n} \mathcal{A}_2(\mathcal{W})^*$ and to $\mathcal{A}_2(\mathcal{W})^*$, respectively.

---

[1] The Grassmannian $\mathrm{Grass}(V, r)$ is isomorphic to the projectivization of the image of the multilinear map

$$\varphi : V^r \dashrightarrow \bigwedge^r V$$

$$(v_1, \ldots, v_r) \mapsto v_1 \wedge \cdots \wedge v_r \,.$$

The isomorphism is given of course by associating with any $W \in \mathrm{Grass}(V, r)$ the point $[\varphi(w_1, \ldots, w_r)] \in \mathbb{P}(\wedge^r V)$ where $w_1, \ldots, w_r$ is an arbitrary basis of W. Clearly, $[\varphi(w_1, \ldots, w_r)]$ does not depend on the chosen basis. The induced map $\mathrm{Grass}(V, r) \to \mathbb{P}(\wedge^r V)$ is the so-called **Plücker embedding** of $\mathrm{Grass}(V, r)$.

Consequently, the identity

$$\big(\widehat{P_{\mathcal{W}} \circ \gamma}\big)'(t) \wedge (\widehat{\kappa_j} \circ \gamma)(t) = -\big(\widehat{P_{\mathcal{W}} \circ \gamma}\big)(t) \wedge \big(\widehat{\kappa_j} \circ \gamma\big)'(t)$$

holds true for every $j \in B$.

Lemma 5.2.3 ensures the non-vanishing of this latter expression. This implies in particular that $\big(\widehat{P_{\mathcal{W}} \circ \gamma}\big)'(t)$ does not vanish. Since γ is an arbitrary immersion, the differential of $P_{\mathcal{W}}$ is injective at the origin. The proposition follows. □

Corollary 5.2.5. *For every pair of distinct points* $x, y \in (\mathbb{C}^n, 0)$, *the intersection* $\mathbb{P}^{k-n-1}(x) \cap \mathbb{P}^{k-n-1}(y)$ *is a projective subspace of* $\mathbb{P}\mathcal{A}_2(\mathcal{W})^*$ *of dimension* $n - 2$.

Proof. It suffices to prove the claim for $x = 0$ and y arbitrarily close to it. Since $2(k - n - 1) - (2k - 3n) = n - 2$, the claim is equivalent to the transversality of $\mathbb{P}^{k-n-1}(0)$ and $\mathbb{P}^{k-n-1}(y)$. The reader is invited to verify that a non-trivial lack of transversality between $\mathbb{P}^{k-n-1}(0)$ and $\mathbb{P}^{k-n-1}(y)$, for y arbitrarily close to 0, would imply that the differential of $P_{\mathcal{W}}$ at the origin is not injective. This contradiction implies the corollary. □

5.3 Poincaré–Blaschke's Map

In this section the Poincaré–Blaschke's map of a web satisfying the assumptions of Theorem 5.1.1 is defined and it is proved that it has rank 2 when the ambient dimension is at least 3. Its content is considerably more technical, although rather elementary, than the remaining of the book. The arguments herein follow very closely [130].

Notation

Let $u_1, \ldots, u_k : (\mathbb{C}^n, 0) \to (\mathbb{C}, 0)$ be germs of holomorphic submersions defining the foliations $\mathcal{F}_1, \ldots, \mathcal{F}_k$. Recall that Proposition 2.4.10 establishes the existence of a coframe $\varpi = (\varpi_0, \ldots, \varpi_{n-1})$ for $\Omega^1(\mathbb{C}^n, 0)$, and k holomorphic functions $\theta_1, \ldots, \theta_k$, such that the foliation $\mathcal{F}_i$ is induced by the 1-form

$$\omega_i = \sum_{q=0}^{n-1} (\theta_i)^q \varpi_q. \tag{5.3}$$

Since $\mathcal{W}$ is smooth, one has $\theta_i(x) \neq \theta_j(x)$ for every distinct $i, j \in \underline{k}$ and for every $x \in (\mathbb{C}^n, 0)$. Notice also the existence of non-vanishing holomorphic functions $h_1, \ldots, h_k$ satisfying $d u_i = h_i \omega_i$ for every $i \in \underline{k}$.

Until the end of Sect. 5.3, the submersions u_i, the coframe ϖ, the functions θ_i and h_i, and the 1-forms ω_i will have the same meaning as above.

For an arbitrary 1-form $\alpha \in \Omega^1(\mathbb{C}^n, 0)$ its qth component in the coframe ϖ will be written as $\{\alpha\}^q$. More precisely, the holomorphic functions $\{\alpha\}^0, \ldots, \{\alpha\}^{n-1}$ are implicitly defined by the identity

$$\alpha = \sum_{q=0}^{n-1} \{\alpha\}^q \varpi_q .$$

To write down the canonical maps κ_i explicitly, identify the point $[a_1 : \cdots : a_\pi] \in \mathbb{P}^{\pi-1} \simeq \mathbb{P}\mathcal{A}_2(\mathcal{W})$ with the hyperplane $\{a_1 \eta^{(1)} + \cdots + a_\pi \eta^{(\pi)} = 0\}$ in[2] $\mathcal{A}_2(\mathcal{W})$. The ith evaluation morphism at a point $x \in (\mathbb{C}^n, 0)$ is nothing else than

$$(a_1, \ldots, a_\pi) \longmapsto a_1 \eta_i^{(1)}(x) + \cdots + a_\pi \eta_i^{(\pi)}(x) .$$

Notice that the 1-forms $\eta_i^{(j)}$ for $j \in \underline{\pi}$, are all proportional to ω_i. Hence there are holomorphic functions $z_i^{(j)}$ such that

$$\eta_i^{(j)} = z_i^{(j)} \omega_i$$

for every $i \in \underline{k}$ and every $j \in \underline{\pi}$. Therefore, for a fixed $i \in \underline{k}$, the map

$$Z_i : (\mathbb{C}^n, 0) \longrightarrow \mathbb{C}^\pi$$

$$x \longmapsto \left(z_i^{(1)}(x), \ldots, z_i^{(\pi)}(x) \right)$$

is a lift of κ_i to $\mathbb{C}^\pi$. More precisely, the diagram

$$
\begin{array}{ccc}
 & & \mathbb{C}^\pi \simeq \mathcal{A}_2(\mathcal{W})^* \\
 & \overset{Z_i}{\nearrow} & \Big| \\
(\mathbb{C}^n, 0) & \overset{\kappa_i}{\longrightarrow} & \mathbb{P}\mathcal{A}_2(\mathcal{W})^*
\end{array}
$$

commutes.

For further use, the translation of the conditions $\sum \eta_i = 0$ and $d\eta_i = 0$ to conditions on the functions $z_i^{(j)}$ is stated below as a lemma. The proof is immediate.

---

[2]Here the abelian relations $\eta^{(j)}$ are seen as coordinate functions on $\mathcal{A}_2(\mathcal{W})$, which is the same as thinking of them as elements of $\mathcal{A}_2(\mathcal{W})^*$.

Lemma 5.3.1. *If $z_1, \ldots, z_k : (\mathbb{C}^n, 0) \to \mathbb{C}$ are holomorphic functions on $(\mathbb{C}^n, 0)$, then $(z_1 \omega_1, \ldots, z_k \omega_k)$ is an abelian relation of $\mathcal{W}$ if and only if*

$$d(z_i \, \omega_i) = 0 \qquad and \qquad \sum_{i=1}^{k} z_i \, (\theta_i)^{\sigma} = 0 \qquad (5.4)$$

for every $i \in \underline{k}$ and every $\sigma \in \{0, \ldots, n-1\}$.

5.3.1 Interpolation of the Canonical Maps

For $i \in \underline{k}$, let us consider the polynomial $P_i \in \mathcal{O}(\mathbb{C}^n, 0)[t]$ defined through the formula

$$P_i(t) = \prod_{\substack{j=1 \\ j \neq i}}^{k} (t - \theta_j).$$

Up to multiplication by non-vanishing numbers, the canonical maps, or more precisely their lifts Z_i defined above, are interpolated by the map Z_* defined below:

$$Z_* : (\mathbb{C}^n, 0) \times \mathbb{C} \longrightarrow \mathbb{C}^\pi$$

$$(x, t) \longmapsto \sum_{i=1}^{k} P_i(t) Z_i(x).$$

Indeed $Z_*(x, \theta_i(x))$ is proportional to $Z_i(x)$ since

$$Z_*(x, \theta_i(x)) = P_i(\theta_i(x)) Z_i(x) \quad \text{with} \quad P_i(\theta_i(x)) \neq 0.$$

Some properties of Z_* are collected in the following lemma.

Lemma 5.3.2. *The map Z_* has the following properties:*

(a) for every $x \in (\mathbb{C}, 0)$ and every $t \in \mathbb{C}$, $Z_*(x, t) \neq 0$;
(b) the entries of $Z_*(x, t)$, seen as polynomials in $\mathcal{O}_{(\mathbb{C}^n, 0)}[t]$, have degree at most $k - n - 1$;
(c) the coefficient of t^{k-n-1} in $Z_*(x, t)$ is non-zero and equal to $\sum_{i=1}^{k} \theta_i(x)^n Z_i(x)$.

Proof. Let (x_0, t_0) be a point of $(\mathbb{C}^n, 0) \times \mathbb{C}$. If $t_0 = \theta_i(x_0)$ for some $i \in \underline{k}$, then clearly $Z_*(x_0, t_0) = P_i(\theta_i(x_0)) Z_i(x_0) \neq 0$.

Assume now that t_0 belongs to $\mathbb{C} \setminus \{\theta_1(x_0), \ldots, \theta_k(x_0)\}$. Because $\ell^1(\mathcal{W}) = k - n$, Lemma 5.3.1 implies that every relation of the form $\sum_i c_i Z_i(x_0) = 0$ is a linear combination of the relations $\sum_i \theta_i(x_0)^\sigma Z_i(x_0)$ for $\sigma = 0, \ldots, n - 1$. If

$Z_*(x_0, t_0) = 0$, then $\sum_i (t_0 - \theta_i(x_0))^{-1} Z_i(x_0)$ would be equal to 0 and the preceding argument would imply that there exist some constants $\mu_1, \ldots, \mu_n \in \mathbb{C}$ which satisfy

$$\frac{1}{t_0 - \theta_i(x_0)} = \sum_{\sigma=1}^{n} \mu_\sigma \, \theta_i(x_0)^\sigma$$

for every $i \in \underline{k}$. But this is not possible, since $\theta_i(x_0) \neq \theta_j(x_0)$ whenever $i \neq j$. Hence item (a) holds true.

To prove item (b), the dependence on x will be dropped from the notation in order to keep it simple. Let $P(t) = \prod_{j=1}^{k}(t - \theta_j)$ and write

$$P_i(t) = \sum_{j=0}^{k-1} \sigma_j^{(i)} t^j \quad \text{and} \quad P(t) = \sum_{j=0}^{k} \sigma_j t^j \tag{5.5}$$

for some holomorphic coefficient $\sigma_j^{(i)}, \sigma_j$ depending only on $x \in (\mathbb{C}^n, 0)$.

Comparing coefficients in the identities $P(t) = (t - \theta_i) P_i(t)$, one gets the following relations for every $i \in \underline{k}$:

$$\sigma_j = \begin{cases} - \theta_i \, \sigma_0^{(i)} & \text{if } j = 0; \\ \sigma_{j-1}^{(i)} - \theta_i \, \sigma_j^{(i)} & \text{if } j = 1, \ldots, k-1; \\ \sigma_{k-1}^{(i)} & \text{if } j = k. \end{cases}$$

Notice that $\sigma_k = \sigma_{k-1}^{(i)} = 1$ for every i since $P(t)$ and $P_i(t)$ are monic polynomials. Consequently, for every $i \in \underline{k}$ and every $j = 0, \ldots, k-1$, one has

$$\sigma_j^{(i)} = \sum_{s=0}^{k-1-j} (\theta_i)^s \, \sigma_{j+1+s} . \tag{5.6}$$

From Eq. (5.6), it follows that

$$Z_*(t) = \sum_{j=0}^{k-1} \left(\sum_{i=1}^{k} \sigma_j^{(i)} Z_i \right) t^j$$

$$= \sum_{j=0}^{k-1} \left(\sum_{i=1}^{k} \sum_{s=0}^{k-j-1} (\theta_i)^s \, \sigma_{j+s+1} Z_i \right) t^j$$

$$= \sum_{j=0}^{k-1} \left[\sum_{s=0}^{k-j-1} \left(\sum_{i=1}^{k} (\theta_i)^s Z_i \right) \sigma_{j+s+1} \right] t^j . \tag{5.7}$$

According to Lemma 5.3.1, $\sum_i Z_i(\theta_i)^s$ is identically zero for any $s \in \{0, \ldots, n-1\}$. Thus the coefficient of t^j in (5.7) is identically zero for every $j \geq k - n$. Item (b) follows.

The coefficient of t^{k-n-1} in $Z_*(t)$ is

$$\sum_{s=0}^{n} \left(\sum_{i=1}^{k} (\theta_i)^s Z_i \right) \sigma_{(k-n-1)+s+1} = \left(\sum_{i=1}^{k} (\theta_i)^n Z_i \right) \sigma_k ,$$

the equality being obtained through the use of Lemma 5.3.1 exactly as above. One proves that $\sum_{i=1}^{k} (\theta_i)^n Z_i \neq 0$ by using the very same arguments used to establish Item (a). Since $\sigma_k = 1$, Item (c) follows. $\qquad\square$

5.3.2 Definition of Poincaré–Blaschke's Map

The Poincaré–Blaschke's map of $\mathcal{W}$,

$$PB_{\mathcal{W}} : (\mathbb{C}^n, 0) \times \mathbb{P}^1 \longrightarrow \mathbb{P}^{\pi-1} ,$$

is defined as

$$PB_{\mathcal{W}}(x,t) = \begin{cases} \left[Z_*(x,t) \right] & \text{for } t \in \mathbb{C} \\ \left[\sum_{i=1}^{k} \theta_i(x)^n Z_i(x) \right] & \text{for } t = \infty. \end{cases}$$

The name comes from the fact that such a map was first introduced by Blaschke extrapolating ideas of Poincaré [117]. In [16], Blaschke constructed and studied the Poincaré–Blaschke map of a maximal rank planar 5-web. He mistakenly asserted that its image lies in a surface of $\mathbb{P}^5$.

Observe that the definition of Z_* does depend on the choice of the subspace $\mathcal{A}_2(\mathcal{W}) \subset \mathcal{A}(\mathcal{W})$; on the basis of $\mathcal{A}_2(\mathcal{W})$; and on the adapted coframe $(\varpi_0, \ldots, \varpi_{n-1})$. Nevertheless, modulo projective changes of coordinates on the target $\mathbb{P}^{\pi-1}$ and on the $\mathbb{P}^1$ factor of the source, $PB_{\mathcal{W}}$ only depends on the choice of the subspace $\mathcal{A}_2(\mathcal{W}) \subset \mathcal{A}(\mathcal{W})$ as the reader can easily verify.

Notice that the Poincaré–Blaschke's map restricted to $\{x\} \times \mathbb{P}^1$ parametrizes a rational curve which interpolates the *canonical points* $\kappa_1(x), \ldots, \kappa_k(x)$. More precisely,

Lemma 5.3.3. *For $x \in (\mathbb{C}^n, 0)$ fixed, the map*

$$\varphi_x : t \in \mathbb{P}^1 \mapsto PB_{\mathcal{W}}(x,t) \in \mathbb{P}^{\pi-1}$$

is an isomorphism onto a rational normal curve $\mathscr{C}(x) \subset \mathbb{P}^{k-n-1}(x)$ of degree $k - n - 1$.

Proof. Clearly the map under study parametrizes a rational curve $\mathscr{C}(x)$. Moreover $\varphi_x(\theta_i(x)) = [Z_i(x)] = \kappa_i(x)$ for every $i \in \underline{k}$, and

$$\dim \langle \mathscr{C}(x) \rangle \geq \dim \Big\langle Z_1(x), \ldots, Z_k(x) \Big\rangle - 1 = k - n - 1,$$

since the span of $Z_1(x), \ldots, Z_k(x)$ has dimension $\ell^1(\mathcal{W}) = k - n$. Consequently, one has

$$\langle \mathscr{C}(x) \rangle = \mathbb{P}^{k-n-1}(x).$$

Moreover, Lemma 5.3.2 tells that the map Z_* has degree $k - n - 1$ in t. Hence φ_x parametrizes a non-degenerate rational curve in $\mathbb{P}^{k-n-1}(x)$ of degree at most $k - n - 1$. It follows from Proposition 2.4.11 that $\mathscr{C}(x)$ is a rational normal curve of degree $k - n - 1$. $\qquad\square$

5.3.3   The Rank of Poincaré–Blaschke's Map

This section is devoted to the proof of the following proposition.

Proposition 5.3.4. *If $n > 2$, then $PB_{\mathcal{W}}$ has rank 2 at every point of $(\mathbb{C}^n, 0) \times \mathbb{P}^1$.*

We will first deal rather easily with the case of some specific points of $(\mathbb{C}^n, 0) \times \mathbb{P}^1$. The case of the other ones will be treated next. The latter case is considerably more involved than the former and will necessitate to deal with the most technical arguments of this book.

For $i \in \underline{k}$, let Ξ_i be the graph of θ_i et denote by $\Xi_{\mathcal{W}}$ their union:

$$\Xi_i = \Big\{ (x, t) \in (\mathbb{C}^n, 0) \times \mathbb{P}^1 \mid \theta_i(x) = t \Big\} \quad \text{and} \quad \Xi_{\mathcal{W}} = \bigcup_{i=1}^{k} \Xi_i .$$

The Rank of $PB_{\mathcal{W}}$ at Points of $\Xi_{\mathcal{W}}$

Using some of the preceding results, there is no difficulty to determine the rank of $PB_{\mathcal{W}}$ at any point of $\Xi_{\mathcal{W}}$:

Lemma 5.3.5. *For every $(x, t) \in \Xi_{\mathcal{W}}$, the rank of $PB_{\mathcal{W}}$ at (x, t) is two.*

Proof. According to Lemma 5.3.3, $PB_{\mathcal{W}}$ restricted to the line $\{x\} \times \mathbb{P}^1$ is an isomorphism onto the rational normal curve $\mathscr{C}(x) \subset \mathbb{P}^{k-n-1}(x)$. In particular, the tangent line to $\mathscr{C}(x)$ at $PB_{\mathcal{W}}(x, t)$, namely

$$\Big\langle PB_{\mathcal{W}}(x, t), \frac{\partial PB_{\mathcal{W}}}{\partial t}(x, t) \Big\rangle$$

is contained in $\mathbb{P}^{k-n-1}(x)$.

It follows from Lemma 5.2.1 that the restriction of $PB_{\mathcal{W}}$ to the hypersurface Ξ_i is a submersion onto the ith canonical curve of C_i. Thus the image of the differential of $PB_{\mathcal{W}}|_{\Xi_i}$ at $(x, \theta_i(x))$ is $T_{\kappa_i(x)}C_i$. Lemma 5.2.3 implies that the rank of $PB_{\mathcal{W}}$ is exactly two at any point of the hypersurface Ξ_i. $\qquad\square$

A Holomorphic Lift for the Poincaré–Blaschke's Map Outside $\Xi_{\mathcal{W}}$

Since any term of the form $t - \theta_i(x)$ does not vanish when $(x, t) \notin \Xi_{\mathcal{W}}$, the map

$$Z : (\mathbb{C}^n, 0) \times \mathbb{P}^1 \setminus \Xi_{\mathcal{W}} \longrightarrow \mathbb{C}^\pi$$

$$(x, t) \longmapsto \sum_{i=1}^{k} \frac{Z_i(x)}{t - \theta_i(x)}$$

is well defined and holomorphic. Since $Z_*(x, t) = P(x, t) \cdot Z(x, t)$ for every (x, t) outside $\Xi_{\mathcal{W}}$, it follows that Z is also a lift of the Poincaré–Blaschke's map $PB_{\mathcal{W}}$ of $\mathcal{W}$ to $\mathbb{C}^\pi$.

The following conventions will be used in what follows: the usual exterior derivative on $(\mathbb{C}^n, 0) \times \mathbb{P}^1$ will be denoted by $\underline{d}$, while the exterior differential on $(\mathbb{C}^n, 0)$ will be denoted by d. To clarify:

$$\underline{d}F = dF + \left(\frac{\partial F}{\partial t}\right)dt$$

for every germ of holomorphic function F on $(\mathbb{C}^n, 0) \times \mathbb{P}^1$.

For further reference, observe that the holomorphic $\mathbb{C}^\pi$-valued form dZ can be written as

$$dZ = \sum_{i=1}^{k}(t - \theta_i)^{-1}dZ_i + \sum_{i=1}^{k}(t - \theta_i)^{-2}Z_i\,d\theta_i. \tag{5.8}$$

One defines a holomorphic 1-form on $(\mathbb{C}^n, 0) \times \mathbb{P}^1$ by setting

$$\Pi(x, t) = \begin{cases} \sum_{p=0}^{n-1} t^p \varpi_p(x) & \text{if } t \neq \infty \\ \omega_{n-1}(x) & \text{if } t = \infty. \end{cases}$$

The main technical result used to establish Proposition 5.3.4 is the following one:

Proposition 5.3.6. *There are holomorphic 1-forms Ω and Γ on $(\mathbb{C}^n, 0) \times \mathbb{P}^1$ such that*

$$\underline{d}Z = \{dZ\}^0\,\Pi + Z\,\Omega + (\partial Z/\partial t)\,(\Gamma + dt). \tag{5.9}$$

To prove that the total differential of Z takes the form above is delicate as the next few pages will testify.

The Main Technical Point

The next result is essential in the proof of Proposition 5.3.6.

Lemma 5.3.7. *There are germs of holomorphic functions $M_r^p \in \mathcal{O}(\mathbb{C}^n, 0)$ such that for any abelian relation $(z_1\omega_1, \ldots, z_k\omega_k) \in \mathcal{A}(\mathcal{W})$, for every $p \in \{0, \ldots, n-2\}$ and every $i \in \underline{k}$, the following identity holds true*

$$\{dz_i\}^{p+1} - \theta_i\{dz_i\}^p - z_i\{d\theta_i\}^p = z_i \sum_{r=0}^{n-1}(\theta_i)^r M_r^p. \tag{5.10}$$

Moreover, if $n > 2$, then

$$\{d\theta_i\}^{p+1} - \theta_i\{d\theta_i\}^p = \sum_{\rho=0}^{n}(\theta_i)^\rho N_\rho^p, \tag{5.11}$$

where N_r^p are holomorphic functions on $(\mathbb{C}^n, 0)$.

Remark 5.3.8. Let θ be any function on $(\mathbb{C}^n, 0)$ and set $\omega_\theta = \sum_q \theta^q \varpi_q$. If ω_θ is integrable then, after writing down the coefficients of $\varpi_p \wedge \varpi_p \wedge \varpi_r$ in $\omega_\theta \wedge d\omega_\theta$ and imposing their vanishing, one obtains relations of the form

$$\{d\theta\}^{p+1} - \theta\{d\theta\}^p = \sum_{\rho=0}^{n+p+1} \theta^\rho \mathcal{N}_\rho^p \tag{5.12}$$

for $p = 0, \ldots, n - 2$, where $\mathcal{N}_\rho^p$ are certain holomorphic functions that do not depend on θ but only on the adapted coframe ϖ.

Similarly, one can prove that there are holomorphic functions $\mathcal{M}_\rho^p$ depending only on ϖ such that if ω_θ is integrable, then any z such that $d(z\omega_\theta) = 0$ necessarily verifies

$$\{dz\}^{p+1} - \theta\{dz\}^p - z\{d\theta\}^p = z\sum_{\rho=0}^{n+p} \theta^\rho \mathcal{M}_\rho^p \tag{5.13}$$

for any $p = 0, \ldots, n - 2$. Relations (5.12) and (5.13) are direct consequences of the integrability condition and have nothing to do with webs and/or their abelian relations. Lemma 5.3.7 improves these relations by lowering the upper limit of both summations.

Proof of the Main Technical Point I: Preliminaries

For every $i \in \underline{k}$, the Taylor expansion of u_i centered at the origin writes

$$u_i(x) = \ell_i(x) + \frac{1}{2} q_i(x) + O_0(3)$$

where ℓ_i (resp. q_i) is a linear (resp. a quadratic) form. Let $\xi = (\xi_1 du_1, \ldots, \xi_k du_k)$ be an abelian relation of $\mathcal{W}$. Since $d\xi_i \wedge du_i(0) = 0$, the following identity holds

$$\xi_i(x) = a_i + b_i \ell_i(x) + O_0(2) \qquad i \in \underline{k}$$

for suitable complex numbers a_i, b_i. Looking at the order one jet at the origin of the relation $\sum_i \xi_i du_i = 0$, one deduces that

$$\sum_{i=1}^{k} a_i \ell_i = 0 \quad \text{and} \quad \sum_{i=1}^{k} a_i q_i + \sum_{i=1}^{k} b_i (\ell_i)^2 = 0. \tag{5.14}$$

Let us denote by $\mathcal{Q} = \mathbb{C}_2[x_1, \ldots, x_n]$ the space of quadratic forms on $\mathbb{C}^n$. For $Q = \sum_{i \leq j} Q^{ij} x_i x_j \in \mathcal{Q}$ define the following differential operators

$$Q(\nabla F) = \sum_{i \leq j} Q^{ij} \frac{\partial F}{\partial x_i} \frac{\partial F}{\partial x_j} \quad \text{and} \quad Q_\partial(F) = \sum_{i \leq j} Q^{ij} \frac{\partial^2 F}{\partial x_i \partial x_j}$$

where F is a germ of holomorphic function.

By hypothesis $F^0 \mathcal{A}(\mathcal{W}) / F^1 \mathcal{A}(\mathcal{W})$ has dimension $k - n$. Therefore for every $a = (a_1, \ldots, a_d) \in \mathbb{C}^d$, the following implication holds true

$$\sum_{i=1}^{k} a_i \ell_i = 0 \Longrightarrow \sum_{i=1}^{k} a_i q_i \in \mathrm{Span}_{\mathbb{C}} \left\langle (\ell_1)^2, \ldots, (\ell_k)^2 \right\rangle. \tag{5.15}$$

To better understand this relation, notice that the smoothness of $\mathcal{W}$ implies that $\ell_1, \ldots, \ell_n$ is a basis of $\mathbb{C}_1[x_1, \ldots, x_n]$. Thus for every $i \in \underline{k}$, there is a decomposition $\ell_i = \sum_{j=1}^{n} l_i^j \ell_j$ with constants l_i^j uniquely determined. Thus (5.15) translates into to the more precise statement

$$q_i - \sum_{j=1}^{n} l_i^j q_j \in \mathrm{Span}_{\mathbb{C}} \left\langle (\ell_1)^2, \ldots, (\ell_k)^2 \right\rangle \tag{5.16}$$

for any $i \in \underline{k}$.

If $G \in \mathcal{Q}$ and ℓ is a linear form, then $G_\partial(\ell^2) = 2G(\nabla \ell)$. Suppose that $G(\nabla \ell_i) = 0$ for every $i \in \underline{k}$. Thus $G_\partial(q) = 0$ for every $q \in \mathrm{Span}_{\mathbb{C}} \left\langle (\ell_1)^2, \ldots, (\ell_k)^2 \right\rangle$.

Using the relations (5.16) one deduces

$$G_\partial(q_i) = \sum_{j=1}^{n} l_i^j \, G_\partial(q_i)$$

for every $i \in \underline{k}$.

Since G_∂ is a homogeneous linear operator of order two, one has $G_\partial(\ell) = 0$ for every linear form ℓ, thus

$$G_\partial(u_i)(0) = G_\partial\left(\ell_i + \frac{1}{2}q_i + O_0(3)\right)(0)$$

$$= \frac{1}{2}G_\partial(q_i) + G_\partial(O_0(3))(0) = \frac{1}{2}G_\partial(q_i).$$

Set X_G as the vector field $\frac{1}{2}\sum_{j=1}^{n} G_\partial(q_j)\frac{\partial}{\partial x_j}$. From the preceding relation, one deduces that for every $i \in \underline{k}$, one has

$$G_\partial(u_i)(0) = \langle d\,u_i(0), X_G\rangle.$$

By hypothesis, the implication (5.15) holds true for every $x \in (\mathbb{C}^n, 0)$. The discussion above implies the following result.

Lemma 5.3.9. *Let* $\mathcal{G} = \sum_{i \le j} \mathcal{G}^{ij}(x)x_i x_j$ *be a field of quadratic forms. If* $\mathcal{G}(\nabla u_i)$ *vanishes identically for every* $i \in \underline{k}$, *then there exists a vector field* $X_\mathcal{G}$ *such that for every* $i \in \underline{k}$:

$$\mathcal{G}_\partial(u_i) = \langle d\,u_i, X_\mathcal{G}\rangle.$$

Proof of the Main Technical Point II: Conclusion

Let $(x_1, \ldots, x_n)$ be the standard system of coordinates on $\mathbb{C}^n$. Since both $(dx_1, \ldots, dx_n)$ and $\varpi = (\varpi_0, \ldots, \varpi_{n-1})$ are coframes on $(\mathbb{C}^n, 0)$, there are basis change formulas (for $j = 1, \ldots, n$ and $q = 0, \ldots, n - 1$)

$$dx_j = \sum_{q=0}^{n-1} B_j^q \, \varpi_q \qquad \text{and} \qquad \varpi_q = \sum_{j=1}^{n} C_q^j \, dx_j.$$

For $l, m \in \underline{n}$, set for every $i \in \underline{k}$:

$$u_{i,l} = \partial u_i / \partial x_l \qquad \text{and} \qquad u_{i,lm} = \partial^2 u_i / \partial x_l \partial x_m.$$

Thus

$$du_i = h_i \sum_{q=0}^{n-1} (\theta_i)^q \varpi_q = \sum_{j=1}^{n} u_{i,j}\, dx_j$$

and consequently (for $p = 0, \ldots, n-1$ and $j = 1, \ldots, n$)

$$h_i(\theta_i)^p = \sum_{j=1}^{n} B_j^p\, u_{i,j} \qquad \text{and} \qquad u_{i,j} = h_i \sum_{p=0}^{n-1} C_p^j (\theta_i)^p. \tag{5.17}$$

Let us consider four integers p, p', q, q' in the interval $[0, n-1]$ which satisfy $p + q = p' + q'$. Because $(\theta_i)^p (\theta_i)^q = (\theta_i)^{p'} (\theta_i)^{q'}$, the Eq. (5.17) imply that the relation

$$\sum_{j,j'=1}^{n} \left(B_j^p B_{j'}^q - B_j^{p'} B_{j'}^{q'} \right) u_{i,j}\, u_{i,j'} = 0$$

holds true for every $i \in \underline{k}$.

Lemma 5.3.9 implies the existence of holomorphic functions $X^1, \ldots, X^n$, for which

$$\sum_{j,j'=1}^{n} \left(B_j^p B_{j'}^q - B_j^{p'} B_{j'}^{q'} \right) u_{i,jj'} = \sum_{l=1}^{n} X^l u_{i,l}$$

for every $i \in \underline{k}$.

Notice that

$$d\left(h_i(\theta_i)^p \right) = \sum_{j=1}^{n} \partial_{x_j} \left(h_i(\theta_i)^p \right) dx_j = \sum_{q=0}^{n-1} \left[\sum_{j=1}^{n} B_j^q \partial_{x_j} \left(h_i(\theta_i)^p \right) \right] \varpi_q$$

and consequently $\{ d(h_i(\theta_i)^p) \}^q = \sum_j B_j^q \partial_{x_j} (h_i(\theta_i)^p)$.

Combining this last equation with (5.17) one obtains

$$\{ d(h_i(\theta_i)^p) \}^q = \sum_{j,j'=1}^{n} B_j^q \partial_{x_j} (B_{j'}^p u_{i,j'})$$

$$= \sum_{j,j'=1}^{n} B_j^q \partial_{x_j} (B_{j'}^p) u_{i,j'} + \sum_{j,j'=1}^{n} B_j^q B_{j'}^p u_{i,jj'}.$$

Thus, one can write

$$\{d(h_i(\theta_i)^q)\}^p - \{d(h_i(\theta_i)^{q'})\}^{p'} = \sum_{l=1}^{n} Y^l u_{i,l}$$

with

$$Y^l = X^l + \sum_{j,j'=1}^{n} \left(B_{j'}^p \partial_{x_j}(B_l^q) - B_{j'}^{p'} \partial_{x_j}(B_l^{q'})\right).$$

Once again one has to apply the relations (5.17). After setting $M_r^{pq} = \sum_l Y^l C_r^l$ for $r = 0, \ldots, n-1$, it follows that

$$\{d(h_i(\theta_i)^q)\}^p - \{d(h_i(\theta_i)^{q'})\}^{p'} = h_i \sum_{r=0}^{n-1} M_r^{pq} (\theta_i)^r \qquad (5.18)$$

for no matter which $i \in \underline{k}$.

Note that the functions M_r^{pq} depend only on the integers p, q, p', q', but not on i. It suffices to take $p' = p+1$, $q = 1$ and $q' = 0$ to establish the existence of functions M_r^p satisfying

$$\{d(h_i\theta_i)\}^p - \{dh_i\}^{p+1} = h_i \sum_{r=0}^{n-1} M_r^p (\theta_i)^r. \qquad (5.19)$$

Let $z = (z_1\omega_1, \ldots, z_k\omega_k)$ be an abelian relation of $\mathcal{W}$. The function z_i is such that $d(z_i\omega_i) = 0$ then $d(z_i/h_i) \wedge du_i = 0$. The latter relation writes

$$\left(\frac{dz_i}{h_i} - \frac{z_i dh_i}{h_i^2}\right) \wedge h_i\omega_i = \left(dz_i - \frac{z_i dh_i}{h_i}\right) \wedge \omega_i = 0.$$

Decomposing it in the coframe ϖ, one obtains

$$\sum_{p=0}^{n-1} \left(\{dz_i\}^p - z_i h_i^{-1}\{dh_i\}^p\right)\varpi_p \wedge \sum_{q=0}^{n-1}(\theta_i)^q \varpi_q = 0,$$

which implies

$$\left(\{dz_i\}^p - z_i h_i^{-1}\{dh_i\}^p\right)(\theta_i)^q - \left(\{dz_i\}^q - z_i h_i^{-1}\{dh_i\}^q\right)(\theta_i)^p = 0 \qquad (5.20)$$

for every $p, q = 0, \ldots, n - 1$. Taking $q = p + 1$ and dividing by $(\theta_i)^p$, one obtains

$$\{dz_i\}^{p+1} - \theta_i\{dz_i\}^p - z_i h_i^{-1}\left(\{dh_i\}^{p+1} - \theta_i\{dh_i\}^p\right) = 0 \,.$$

It suffices to combine it with (5.19) to obtain the relations (5.10) of Lemma 5.3.7.

To obtain the relations (5.11), take $q = 2$, $q' = 1$ and $p' = p + 1$ in (5.18). Note that this is possible only because n is assumed to be at least 3. On the one hand, it follows from (5.18) that there exist holomorphic functions $L_0, \ldots, L_{n-1}$ which satisfy the following identity for every $i \in \underline{k}$:

$$\{d(h_i(\theta_i)^2)\}^p - \{d(h_i\theta_i)\}^{p+1} = h_i \sum_{r=0}^{n-1} L_r\, (\theta_i)^r \,. \tag{5.21}$$

On the other hand, one has

$$\{d(h_i(\theta_i)^2)\}^p - \{d(h_i\theta_i)\}^{p+1} = \theta_i\left(\{d(h_i\theta_i)\}^p - \{dh_i\}^{p+1}\right)$$
$$+ h_i\left(\theta_i\{d\theta_i\}^p - \{d\theta_i\}^{p+1}\right).$$

Plugging these formulae into (5.21) and using (5.19), one finally obtains (5.11) and concludes in this way the proof of Lemma 5.3.7. □

Remark 5.3.10. Note that the condition $n \geq 3$ is only used at the very end of the proof to obtain the relations (5.21) which imply rather straightforwardly the relations (5.11). Except for these last lines, all the arguments above are valid in dimension two.

Three Technical Lemmata

The following simple lemmata will prove to be useful later.

Lemma 5.3.11. *For $l = 0, \ldots, n$ and $L = 0, \ldots, n+1$, respectively, the following identities hold true:*

$$\sum_{i=1}^{k} \frac{Z_i(\theta_i)^l}{t - \theta_i} = t^l\, Z \quad \text{and} \quad \sum_{i=1}^{k} \frac{Z_i\,(\theta_i)^L}{(t - \theta_i)^2} = L\,t^{L-1}\,Z - t^L\,(\partial Z/\partial t).$$

$$\tag{5.22}$$

Proof. Both identities are proved by induction. Notice that they both are trivially true for $l = 0$ and $L = 0$.

Assume that the first identity holds true for $l < n$, and write

$$\sum_{i=1}^{k} \frac{Z_i (\theta_i)^{l+1}}{(t - \theta_i)} = \sum_{i=1}^{k} \frac{Z_i (\theta_i)^l ((\theta_i - t) + t)}{(t - \theta_i)} = -\sum_{i=1}^{k} Z_i (\theta_i)^l + t \sum_{i=1}^{k} \frac{Z_i (\theta_i)^l}{(t - \theta_i)}.$$

Observe that $\sum_{i=1}^{k} Z_i (\theta_i)^l = 0$ according to Eq. (5.4). Using this observation together with the induction hypothesis, it follows that

$$\sum_{i=1}^{k} (t - \theta_i)^{-1} Z_i \, \theta_i^{l+1} = t^{l+1} Z$$

as requested.

Assume now that the second identity holds true for $L \leq n$. Using the same trick as above, one obtains

$$\sum_{i=1}^{k} \frac{Z_i (\theta_i)^{L+1}}{(t - \theta_i)^2} = \sum_{i=1}^{k} \frac{Z_i (\theta_i)^L}{(t - \theta_i)} + t \sum_{i=1}^{k} \frac{Z_i (\theta_i)^L}{(t - \theta_i)^2}.$$

The first identity implies that the first and of the right-hand side is $t^L Z$. The induction hypothesis implies that the second one is $t \cdot (L \, t^{L-1} Z - t^L (\partial Z / \partial t))$. The lemma follows. $\qquad\square$

Lemma 5.3.12. *If the identities (5.10) and (5.11) of Lemma 5.3.7 hold true (in particular if $n > 2$) then, for $p = 0, \ldots, n - 2$, there are functions $F_p, G_p \in \mathcal{O}(\mathbb{C}^n, 0)[t]$ such that*

$$\{dZ\}^{p+1} - t\{dZ\}^p = F_p Z + G_p (\partial Z / \partial t).$$

Proof. Let p be fixed and notice that Eq. (5.8) implies

$$\{dZ\}^p = \sum_{i=1}^{k} \frac{\{dZ_i\}^p}{t - \theta_i} + \sum_{i=1}^{k} \frac{Z_i \{d\theta_i\}^p}{(t - \theta_i)^2}.$$

Decompose $I_p = \{dZ\}^{p+1} - t\{dZ\}^p$ as $K_p + L_p$, where

$$K_p = \sum_{i=1}^{k} \frac{\{dZ_i\}^{p+1} - t \{dZ_i\}^p}{t - \theta_i} \quad \text{and} \quad L_p = \sum_{i=1}^{k} \frac{Z_i \left(\{d\theta_i\}^{p+1} - t \{d\theta_i\}^p \right)}{(t - \theta_i)^2}.$$

Replacing t by $(t - \theta_i) + \theta_i$ in the numerator of K^p gives

$$K_p = \sum_{i=1}^{k} (t - \theta_i)^{-1} \left(\{dZ_i\}^{p+1} - \theta_i \{dZ_i\}^p \right) + \sum_{i=1}^{k} \{dZ_i\}^p.$$

But $\sum_{i=1}^{k} Z_i = 0$ according to (5.4) and consequently

$$K_p = \sum_{i=1}^{k} \frac{\{dZ_i\}^{p+1} - \theta_i \{dZ_i\}^p}{t - \theta_i}. \tag{5.23}$$

In exactly the same way, one proves that

$$L_p = \sum_{i=1}^{k} \frac{\left(\{d\theta_i\}^{p+1} - \theta_i \{d\theta_i\}^p \right) Z_i}{(t - \theta_i)^2} + \sum_{i=1}^{k} \frac{Z_i \{d\theta_i\}^p}{t - \theta_i}.$$

In what follows, let us write $A \equiv B$ if and only if $A - B$ is equal to $FZ + G(\partial Z/\partial t)$ for suitable $F, G \in \mathcal{O}(\mathbb{C}^n, 0)[t]$. Notice that the lemma is equivalent to $I_p \equiv 0$.

The outcome of Lemma 5.3.7, more specifically Eq. (5.11), implies

$$\sum_{i=1}^{k} \frac{\left(\{d\theta_i\}^{p+1} - \theta_i \{d\theta_i\}^p \right) Z_i}{(t - \theta_i)^2} = \sum_{i=1}^{k} \frac{Z_i}{(t - \theta_i)^2} \left(\sum_{\rho=0}^{n} (\theta_i)^\rho N_\rho^p \right).$$

Lemma 5.3.11 in turn implies

$$\sum_{i=1}^{k} \frac{Z_i}{(t - \theta_i)^2} \left(\sum_{\rho=0}^{n} (\theta_i)^\rho N_\rho^p \right) = \sum_{\rho=0}^{n} N_\rho^p \left(\sum_{i=1}^{k} \frac{Z_i (\theta_i)^\rho}{(t - \theta_i)^2} \right) \equiv 0.$$

Therefore

$$L_p \equiv \sum_{i=1}^{k} \frac{Z_i \{d\theta_i\}^p}{t - \theta_i}. \tag{5.24}$$

Combining (5.23) and (5.24), one obtains

$$I_p \equiv \sum_{i=1}^{k} \frac{\{dZ_i\}^{p+1} - \theta_i \{dZ_i\}^p - Z_i \{d\theta_i\}^p}{t - \theta_i}.$$

Equation (5.10) from Lemma 5.3.7 together with Eq. (5.22) from Lemma 5.3.11 allows to conclude:

$$I_p \equiv \sum_{i=1}^{k} \frac{\left(\sum_{r=0}^{n-1} (\theta_i)^r M_r^p \right) Z_i}{t - \theta_i}$$

$$\equiv \sum_{r=0}^{n-1} M_r^p \left(\sum_{i=1}^{k} \frac{Z_i (\theta_i)^r}{t - \theta_i} \right) \equiv \left(\sum_{r=0}^{n-1} M_r^p t^r \right) Z \equiv 0.$$

$\square$

Proof of Proposition 5.3.6

One just has to prove formula (5.9).

Using the notation $I_q = \{dZ\}^{q+1} - t\{dZ\}^q$ (for $q = 0, \ldots, n-2$) introduced in the proof of Lemma 5.3.12, one can write

$$\{dZ\}^p \equiv t^p \{dZ\}^0 + \sum_{q=0}^{p-1} I_q \, t^{p-q-1}$$

for every $p = 0, \ldots, n-1$. It follows that

$$dZ = \sum_{p=0}^{n-1} \{dZ\}^p \varpi_p = \sum_{p=0}^{n-1} \left(t^p \{dZ\}^0 + \sum_{q=0}^{p-1} I_q \, t^{p-1-q} \right) \varpi_p$$

$$= \{dZ\}^0 \Pi + \sum_{p=0}^{n-1} \sum_{q=0}^{p-1} I_q \, t^{p-1-q} \, \varpi_p.$$

Lemma 5.3.12 says that $I^q \equiv 0$ for every $q = 0, \ldots, n-1$ (see page 139 for the definition of $\equiv$). The existence of two 1-forms Ω and Γ satisfying

$$dZ = \{dZ\}^0 \Pi + Z \, \Omega + (\partial Z / \partial t) \, \Gamma$$

follows. Moreover, the coefficients of the 1-forms Ω and Γ in the basis $(\varpi_0, \ldots, \varpi_{n-1})$ are polynomials in t with holomorphic functions on $(\mathbb{C}^n, 0)$ as coefficients. Since $\underline{d}Z = dZ + (\partial Z / \partial t) \, dt$, Proposition 5.3.6 follows.

Proof of Proposition 5.3.4

The kernel of the differential of the canonical map

$$[\,\cdot\,] : \mathbb{C}^\pi \setminus \{0\} \longrightarrow \mathbb{P}^{\pi-1}$$

at a point z of $\mathbb{C}^\pi$ distinct from the origin is the complex line it generates, *i.e.* $\mathrm{Ker}(d\,[\cdot]_z) = \langle z \rangle$. Then it follows from Proposition 5.3.6 that $PB_{\mathcal{W}}$ has rank at most two at every $(x,t) \in (\mathbb{C}^n, 0) \times \mathbb{P}^1$.

At any $(x,t) \in (\mathbb{C}^n, 0) \times \mathbb{P}^1 \setminus \Xi_W$, the 1-forms Π and $\Gamma + dt$ appearing in formula (5.9) are not collinear. Consequently, the image of the total differential $\underline{d}\,Z$ at any such (x,t) is spanned by the three vectors

$$Z(x,t), \quad (\partial Z/\partial t)(x,t) \quad \text{and} \quad \{dZ(x,t)\}^0.$$

In order to prove Proposition 5.3.4, it suffices then to verify that these are linearly independent. This is a consequence of the following stronger lemma:

Lemma 5.3.13. *For any $(x,t) \notin \Xi_W$:*

1. *the vectors $Z(x,t)$ and $(\partial Z/\partial t)(x,t)$ are not collinear;*
2. *the vector $\{dZ(x,t)\}^0$ does not belong to $\mathbb{C}^{k-n}(x) := \langle Z_1(x), \ldots, Z_k(x) \rangle \subset \mathbb{C}^\pi.$*

Proof. To simplify, we do not indicate systematically the dependency in (x,t) that is fixed once for all in $(\mathbb{C}^n, 0) \times \mathbb{P}^1 \setminus \Xi_W$. We will also assume that t is finite. The case when $t = \infty$ can be treated by similar arguments and is left to the reader as an exercise.

Let us assume that Z and $(\partial Z/\partial t)$ are linearly dependent: since Z does not vanish, there exists a complex constant c such that $\partial_t(Z) = c\,Z$, which means that for any $i \in \underline{k}$, one has

$$\sum_{i=1}^{k} \frac{1 + c(t - \theta_i)}{(t - \theta_i)^2} \, Z_i = 0.$$

On the other hand, it follows from Lemma 5.3.1 that the set of relations

$$\sum_{i=1}^{k} (\theta_i)^\sigma Z_i = 0, \quad \sigma \in \{0, \ldots, n-1\}$$

forms a basis of the space of linear relations between the Z_i's. Consequently, it follows that there exist scalars $q_\sigma \in \mathbb{C}$ such that the following relation holds true for every $i \in \underline{k}$:

$$\frac{1 + c(t - \theta_i)}{(t - \theta_i)^2} = \sum_{\sigma=0}^{n-1} q_\sigma (\theta_i)^\sigma.$$

Since the θ_i's are pairwise distinct and because these are at least $2n + 1 > n + 1$ in number, one verifies that it is not possible. This proves the first part of the lemma.

The proof of the second item of the lemma is similar but a bit more technical.

Assume that $\{dZ\}^0$ belongs to the space spanned by the Z_i's. Given the expression (5.8) for the partial differential dZ of Z, this is equivalent to the fact that

$$\sum_{i=1}^{k} \frac{1}{t - \theta_i} \{dZ_i\}^0 = 0 \mod \mathbb{C}^{k-n}(x). \tag{5.25}$$

One verifies that there is no loss in generality by assuming that the basis $\eta^{(1)}, \ldots, \eta^{(\pi)}$ of $A_2(\mathcal{W})$ has been chosen such that the last $k - 2n + 1$ abelian relations

$$\eta^{(k-n+1)}, \ldots, \eta^{(2k-3n+1)}$$

form a basis of $F_x^1 A_2(\mathcal{W})$. For dimensional reasons, this implies that (the classes of) $\eta^{(1)}, \ldots, \eta^{(k-n)}$ form a basis of $A_2(\mathcal{W})/F_x^1 A_2(\mathcal{W})$.

Then let $\mu : \mathbb{C}^\pi \to \mathbb{C}^{k-2n+1}$ be the linear projection that associates $(\zeta^{k+n+1}, \ldots, \zeta^\pi)$ with $(\zeta^1, \ldots, \zeta^\pi) \in \mathbb{C}^\pi$. One has $\mu(\mathbb{C}^{k-n}(x)) = 0$ hence (5.25) implies that the following relation

$$\sum_{i=1}^{k} \frac{1}{t - \theta_i} \{dz_i\}^0 = 0 \tag{5.26}$$

holds true in $\mathbb{C}^{k-2n+1}$, where z_i stands for $\mu(Z_i)$ for every $i \in \underline{k}$.

For every $i \in \underline{k}$, let ℓ_i be $\omega_i(x)$ considered as a linear form on $\mathbb{C}^n$ (remember that x is fixed). It follows from the relations (5.3) that there exists a system of linear coordinates $(x_0, \ldots, x_{n-1})$ on $\mathbb{C}^n$ such that for every $i \in \underline{k}$:

$$\ell_i = \sum_{p=0}^{n-1} (\theta_i(x))^p x_p. \tag{5.27}$$

With these notations, the Taylor expansion at the first order of z_i at x writes

$$z_i = M_i \ell_i + O(2) \tag{5.28}$$

with $M_i \in \mathbb{C}^{k-2n+1}$ and where $O(2)$ is also vector-valued.

One verifies that the subspace

$$M = \langle M_i \mid i \in \underline{k} \rangle \subset \mathbb{C}^{k-2n+1}$$

spanned by the M_i's identifies to $F_x^1 \mathcal{A}_2(\mathcal{W})$. Thus $M = \mathbb{C}^{k-2n+1}$ for dimensional reasons and consequently, there exist exactly $2n-1$ linearly independent linear relations between the M_i's. On the other hand, the latter verifies $\sum_{i=1}^{k} M_i \, \ell_i \, d\ell_i = 0$. Expanding this relation in the coordinates $x_0, \ldots, x_n$ considered above, it follows from (5.27) that the scalar relations

$$\sum_{i=1}^{k} M_i \, (\theta_i)^\sigma = 0, \qquad \sigma = 0, \ldots, 2n - 2 \tag{5.29}$$

form a basis of the space of the linear relations between the M_i's.

Since (5.26) is equivalent to the relation $\sum_{i=1}^{k} \frac{1}{t-\theta_i} M_i = 0$, one obtains that there exist complex constants $q_0, \ldots, q_{2n-2} \in \mathbb{C}$ such that

$$\frac{1}{t - \theta_i} = \sum_{\sigma=0}^{2n-2} q_\sigma (\theta_i)^\sigma$$

for every $i \in \underline{k}$.

Using the same argument as at the end of the proof of the first item of the lemma, one verifies that it is not possible. The hypothesis that $d\{Z(x,t)\}^0$ belongs to $\mathbb{C}^{k-n}(x)$ leads to a contradiction, which finishes the proof of the second assertion of the lemma. □

5.4 Poincaré–Blaschke's Surface

The **Poincaré–Blaschke's surface** $X_{\mathcal{W}}$ of $\mathcal{W}$ is the image of its Poincaré–Blaschke's map $PB_{\mathcal{W}}$. That is

$$X_{\mathcal{W}} = \mathrm{Im}\big(PB_{\mathcal{W}}\big) \subset \mathbb{P}^{\pi-1}.$$

According to Proposition 5.3.4, it is a germ of smooth complex surface on $(\mathbb{P}^{\pi-1}, \mathscr{C}(0))$. It is clearly non-degenerate. The remarks laid down at the end of Sect. 5.3 imply that $X_{\mathcal{W}}$ is canonically attached to the pair $(\mathcal{W}, \mathcal{A}_2(\mathcal{W}))$ modulo projective transformations.

5.4.1 Rational Normal Curves Everywhere

Notice that $X_{\mathcal{W}}$ contains all the canonical curves C_i of $\mathcal{W}$. It also contains a lot of rational curves according to Lemma 5.3.3: the curves $\mathscr{C}(x)$. Exploiting the geometry of this family of rational curves, it will be possible to prove that X_W is the germification of a rational surface along $\mathscr{C}(0)$.

Lemma 5.4.1. *The following assertions are verified:*

(a) *for every $i \in \underline{k}$, the curve $\mathscr{C}(x)$ intersects the canonical curve C_i transversely at $\kappa_i(x)$;*
(b) *for every $x, y \in (\mathbb{C}^n, 0)$, the curve $\mathscr{C}(x)$ coincides with $\mathscr{C}(y)$ if and only if $x = y$;*
(c) *for every subset $J \subset \underline{k}$ of cardinality n, and every set $\mathcal{P} = \{p_j \in C_j \mid j \in J\}$, there exists a unique $x \in (\mathbb{C}^n, 0)$ such that $\mathscr{C}(x)$ contains $\mathcal{P}$.*

Proof. The first assertion follows from the proof of Lemma 5.3.5. It is also a direct consequence of the expression (5.9) for $\underline{d}Z$.

To prove the third assertion, notice that for every $j \in J$, $\kappa_j^{-1}(p_j)$ is a leaf L_j of $\mathcal{F}_j$. Since W is smooth $\cap_{j \in J} L_j$ must reduce to a point $x \in (\mathbb{C}^n, 0)$. The assertion follows.

The second assertion follows immediately from the third.   □

Lemma 5.4.1 implies that the surface X_W is the union of rational normal curves of degree $k - n - 1$ belonging to the n-dimensional holomorphic family

$$\mathscr{C}_W = \{\mathscr{C}(x)\}_{x \in (\mathbb{C}^n, 0)} .$$

The fact that the rational normal curves $\mathscr{C}(x)$'s depend continuously on $x \in (\mathbb{C}^n, 0)$ has the following consequence that will be crucial in order to linearize W in Sect. 5.4.3 below.

Lemma 5.4.2. *The curves $\mathscr{C}(x)$'s are homologous, i.e. their homology classes in $H_1(X_W, \mathbb{Z})$ coincide.*

5.4.2   Algebraization I: X_W Is Algebraizable

Since X_W is smooth, the intersection number $C \cdot C'$ of two curves C and C' contained in it is well defined.

Lemma 5.4.3. *The intersection number between two elements of $\mathscr{C}_W$ is $n - 1$, i.e.*

$$\mathscr{C}(x) \cdot \mathscr{C}(y) = n - 1$$

for every $x, y \in (\mathbb{C}^n, 0)$.

Proof. Let $B \subset \underline{k}$ be a subset of cardinality $n - 1$. Suppose that y is distinct from x and belongs to $\cap_{i \in B} L_i(x)$, where $L_i(x)$ stands for the leaf of $\mathcal{F}_i$ passing through x for any i. After item (c) of Lemma 5.4.1, the curves $\mathscr{C}(x)$ and $\mathscr{C}(y)$ are distinct. Anyway, they have at least $n - 1$ points in common: $p_i = \kappa_i(x) = \kappa_i(y)$ for $i \in B$. Therefore $\mathscr{C}(x) \cdot \mathscr{C}(y) \geq n - 1$.

If $\mathscr{C}(x) \cdot \mathscr{C}(y) \geq n$, then either there exists $p \notin \{p_i\}_{i \in B}$ such that $p \in \mathscr{C}(x) \cap \mathscr{C}(y)$; or there exists $p \in \{p_i\}_{i \in B}$ for which $\mathscr{C}(x)$ and $\mathscr{C}(y)$ are tangent at p. But n points, or $n - 1$ points and one tangent (at one of the $n - 1$ points) on a rational normal curve span a projective subspace of dimension $n - 1$. This contradicts Corollary 5.2.5.

To conclude it suffices to use the fact that any two curves in $\mathscr{C}_W$ are homologous. $\square$

Proposition 5.4.4. *Let* $(S_0, C) \subset \mathbb{P}^N$ *be a germ of smooth surface along a connected projective curve* $C \subset S_0$. *If* $C^2 > 0$, *then* S_0 *is contained in a projective surface* $S \subset \mathbb{P}^N$.

Proof. Let $\mathbb{C}(S_0)$ be the field of meromorphic functions on S_0. To prove that S_0 is contained in a projective surface, it suffices to show that the transcendence degree of $\mathbb{C}(S_0)$ over $\mathbb{C}$ is two. Clearly, since the restriction at S_0 of any two generic rational functions on $\mathbb{P}^N$ are algebraically independent, it suffices to assume that the polar set of both does not contain S_0 and that their level sets are generically transverse to obtain that $\mathrm{tr}_{\deg}[\mathbb{C}(S_0) : \mathbb{C}] \geq 2$.

The hypothesis $C^2 > 0$ is equivalent to the ampleness of the normal bundle N_{C/S_0}. Thus [69, Theorem 6.7] implies that $\mathrm{tr}_{\deg}[\mathbb{C}(S_0) : \mathbb{C}] \leq 2$.

Alternatively, it is also possible to apply a Theorem of Andreotti: since any representative of S_0 contains a curve of positive self-intersection, it also contains a pseudo-concave open subset. Hence, [5, Théorème 6] implies $\mathrm{tr}_{\deg}[\mathbb{C}(S_0) : \mathbb{C}] \leq 2$. $\square$

The preceding proposition together with Lemma 5.4.3 has the following consequence.

Corollary 5.4.5. *Poincaré–Blaschke's surface* X_W *is contained in an irreducible non-degenerate projective surface* $S_W \subset \mathbb{P}^{\pi-1}$.

Remark 5.4.6. Notice that nothing is said about the smoothness of S_W. A priori, it could even happen that the germ of S_W along $\mathscr{C}(0)$ is singular. Of course, this would happen if and only if the germ of S_W along $\mathscr{C}(0)$ had other irreducible components beside X_W. The only clear thing is that S_W contains the smooth surface X_W.

Remark 5.4.7. Readers who are bewildered by the use of the results of Hartshorne or Andreotti in the proof of Proposition 5.4.4 might feel relieved by knowing that Corollary 5.4.5 can be proved by rather elementary means, which are sketched below.

Let x_0 be an arbitrary point of $\mathscr{C}(0)$ and consider the subset $\mathfrak{X}$ of $\mathrm{Mor}_{k-n-1}(\mathbb{P}^1, \mathbb{P}^{\pi-1})^3$ consisting of morphisms ϕ which map $(\mathbb{P}^1, [0 : 1])$ to

---

[3]This is just the set of morphisms from $\mathbb{P}^1$ to $\mathbb{P}^n$ of degree $k - n - 1$ which can be naturally identified with a Zariski open subset of $\mathbb{P}(\mathbb{C}_{k-n-1}[s, t]^\pi)$.

$(X_\mathcal{W}, x_0)$. Recall that the ring of formal power series in any number variables is Noetherian. If $\mathcal{I} \subset \mathbb{C}[[x_1, \ldots, x_{\pi-1}]]$ is the ideal defining $(X_\mathcal{W}, x_0)$ then, expanding formally $f(\phi(t : 1))$ for every defining equation $f \in \mathcal{I}$ one deduces that $\mathfrak{X}$ is algebraic.

To conclude, one has just to prove that the natural projection from $\mathrm{Mor}_{k-n-1}(\mathbb{P}^1, \mathbb{P}^{\pi-1})$ to $\mathbb{P}^{\pi-1}$—the evaluation morphism—sends $\mathfrak{X}$ onto a surface $S_\mathcal{W}$ of $\mathbb{P}^{\pi-1}$ containing $X_\mathcal{W}$.

5.4.3 Algebraization II: Conclusion

Since the projective surface $S_\mathcal{W}$ can be singular, it will be replaced by one of its desingularizations. It can be assumed that the chosen desingularization contains an isomorphic copy of $(X_\mathcal{W}, \mathscr{C}(0))$. Notice also that every desingularization of a singular projective surface is still projective. To keep the notation simple, this desingularization will still be denoted by $S_\mathcal{W}$.

Proposition 5.4.8. *There are no holomorphic 1-forms on $S_\mathcal{W}$, that is*

$$h^0(S_\mathcal{W}, \Omega^1_{S_\mathcal{W}}) = 0.$$

Proof. Let ξ be a holomorphic 1-form on $S_\mathcal{W}$. If non-zero, then it defines a foliation $\mathcal{F}_\xi$ on $S_\mathcal{W}$. Since smooth rational curves have no holomorphic 1-forms, the pull-back of ξ to $\mathscr{C}(x)$ must vanish for every $x \in (\mathbb{C}^n, 0)$. Therefore these curves are invariant by $\mathcal{F}_\xi$. But a foliation on a surface cannot have an n-dimensional family of pairwise distinct leaves. This contradiction shows that ξ is identically zero. $\square$

Hodge theory implies that $H^1(S_\mathcal{W}, \mathcal{O}_{S_\mathcal{W}})$ is also trivial. Therefore from the long cohomology sequence associated with the exponential sequence

$$0 \to \mathbb{Z} \longrightarrow \mathcal{O}_S \longrightarrow \mathcal{O}_S^* \to 0,$$

one deduces that the Chern class morphism

$$H^1(S_\mathcal{W}, \mathcal{O}_{S_\mathcal{W}}^*) \longrightarrow H^2(S_\mathcal{W}, \mathbb{Z})$$

is injective. Consequently, two projective curves C_1 and C_2 in $S_\mathcal{W}$ are linearly equivalent if and only if they are homologous. The direct implication being obvious, let us explain why the converse holds true.

If C_1 and C_2 are homologous, then the Poincaré duals in $H^1(S_\mathcal{W}, \mathbb{Z})$ of their homology classes coincide. By a classical result of complex geometry (see p. 144 of [63] for instance), this means that the Chern classes of the line-bundles $\mathcal{O}_{S_\mathcal{W}}(C_1)$ and $\mathcal{O}_{S_\mathcal{W}}(C_2)$ are the same. The Chern class morphism being injective in our situation, this implies that these two line-bundles are isomorphic hence that the two considered curves are linearly equivalent.

Remark 5.4.9. The surface $S_{\mathcal{W}}$ is rational. To see it, take $n-1$ points in $\mathscr{C}(0)$, say $\kappa_1(0), \ldots, \kappa_{n-1}(0)$. If L_i is a leaf of the foliation $\mathcal{F}_i$ through the origin, then $\cap_{i \in \underline{n-1}} L_i$ is a curve C in $(\mathbb{C}^n, 0)$. By construction, for every $x \in C$ the curve $\mathscr{C}(x)$ intersects $\mathscr{C}(0)$ at $\kappa_1(0), \ldots, \kappa_{n-1}(0)$. Thus blowing up at these $n-1$ points, one obtains a surface S containing a family parametrized by $(\mathbb{C}, 0)$, of rational curves of self-intersection zero: the strict transforms of $\mathscr{C}(x)$ for $x \in C$. Any two of these curves are linearly equivalent, therefore there exists a non-constant holomorphic map $F : S \dashrightarrow \mathbb{P}^1$ sending all of them to points. Thus S is a rational fibration over $\mathbb{P}^1$, hence a rational surface.

Theorem 5.4.10. *The web $\mathcal{W}$ is linearizable.*

Proof. Recall that any two curves in the family $\mathscr{C}_{\mathcal{W}}$ are homologous by Lemma 5.4.2, and consequently linearly equivalent according to the discussion before Remark 5.4.9. Thus they all belong to the complete linear system

$$\left| \mathscr{C}(0) \right| = \mathbb{P} H^0\big(S_{\mathcal{W}}, \mathcal{O}_{S_{\mathcal{W}}}(\mathscr{C}(0))\big).$$

Tensoring the standard exact sequence

$$0 \to \mathcal{O}_{S_{\mathcal{W}}}\big(-\mathscr{C}(0)\big) \longrightarrow \mathcal{O}_{S_{\mathcal{W}}} \longrightarrow \mathcal{O}_{\mathscr{C}(0)} \to 0$$

by $\mathcal{O}_{S_{\mathcal{W}}}(\mathscr{C}(0))$, one obtains

$$0 \to \mathcal{O}_{S_{\mathcal{W}}} \longrightarrow \mathcal{O}_{S_{\mathcal{W}}}\big(\mathscr{C}(0)\big) \longrightarrow \mathcal{O}_{\mathscr{C}(0)}\big(\mathscr{C}(0)\big) \to 0. \tag{5.30}$$

Notice that

$$\deg \mathcal{O}_{\mathscr{C}(0)}\big(\mathscr{C}(0)\big) = \mathscr{C}(0)^2 = n - 1.$$

Consequently $\mathcal{O}_{\mathscr{C}(0)}(\mathscr{C}(0))) \cong \mathcal{O}_{\mathbb{P}^1}(n-1)$.

Since $H^1(S_{\mathcal{W}}, \mathcal{O}_{S_{\mathcal{W}}}) \simeq H^0(S_{\mathcal{W}}, \Omega^1_{S_{\mathcal{W}}}) = 0$, it follows from (5.30) that

$$h^0\big(S_{\mathcal{W}}, \mathcal{O}_{S_{\mathcal{W}}}(\mathscr{C}(0))\big) = h^0\big(S_{\mathcal{W}}, \mathcal{O}_{S_{\mathcal{W}}}\big) + h^0\big(S_{\mathcal{W}}, \mathcal{O}_{\mathscr{C}(0)}(\mathscr{C}(0))\big)$$

$$= 1 + n.$$

Therefore $\left| \mathscr{C}(0) \right| \simeq \mathbb{P}^n$. According to Lemma 5.4.1 item (b), the map

$$\mathfrak{C} : (\mathbb{C}^n, 0) \longrightarrow \left| \mathscr{C}(0) \right| \simeq \mathbb{P}^n$$

$$x \longmapsto \mathscr{C}(x)$$

is injective. Moreover, there exists a factorization

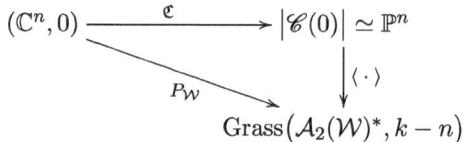

where $\langle \, \cdot \, \rangle$ stands for the map that associates with the curve $\mathscr{C}(x)$ its projective span $\langle \mathscr{C}(x) \rangle$. Since $P_{\mathcal{W}}$ is an immersion (Proposition 5.2.4), so is $\mathfrak{C}$.

The image by $\mathfrak{C}$ of $L_i(x)$, the leaf of $\mathcal{F}_i$ through $x \in (\mathbb{C}^n, 0)$, is nothing else than the germ at $\mathscr{C}(x) = \mathfrak{C}(x)$ of the family of elements of $|\mathscr{C}(0)|$ passing through $\kappa_i(x) \in X_{\mathcal{W}} \subset S_{\mathcal{W}}$. Because $|\mathscr{C}(0)|$ is a linear system, $\mathfrak{C}(L_i(x))$ is a (germ of) hyperplane in $\mathbb{P}^n$. Therefore $\mathfrak{C}$ is a germ of biholomorphism which linearizes $\mathcal{W}$.

$\square$

It suffices to apply the algebraization theorem to the linear webs of maximal rank $\mathfrak{C}_* \mathcal{W}$ to conclude that $\mathcal{W}$ is algebraizable. This concludes the proof of Theorem 5.1.1.

5.5 A Counterexample in Dimension Two

Let $\mathcal{W}$ be a k-web on $(\mathbb{C}^n, 0)$ as above, with $n \geq 2$, the case $n = 2$ being now included. Then the reader will verify that, assuming that $PB_{\mathcal{W}}$ has rank 2 at every point of $(\mathbb{C}^n, 0) \times \mathbb{P}^1$, all the content of the preceding section holds true and consequently, $\mathcal{W}$ is algebraizable.

Conversely, it is not difficult (and left as an exercise to the reader) to verify that if $\mathcal{W}$ is algebraizable, then $PB_{\mathcal{W}}$ is indeed of rank 2 at every point of $(\mathbb{C}^n, 0) \times \mathbb{P}^1$.

Thus the statement of Theorem 5.1.1 can be completed as an equivalence:

Theorem 5.5.1. *Let n and k be integers such that $n \geq 2$ and $k \geq 2n + 1$. Assume that $\mathcal{W}$ is a k-web on $(\mathbb{C}^n, 0)$ such that $\dim F\mathcal{A}(\mathcal{W})/\mathcal{A}_2(\mathcal{W})$ is maximal, equal to $2k - 3n + 1$. Then the following assertions are equivalent:*

- *$\mathcal{W}$ is algebraizable;*
- *$PB_{\mathcal{W}}$ is of rank 2 on $(\mathbb{C}^n, 0) \times \mathbb{P}^1$.*

The preceding result offers an effective way to verify if a given planar web of maximal rank is algebraizable or not. Let us illustrate this with a very simple explicit example.

The following polynomial identities

$$(x) + (y) - (x + y) = 0,$$
$$(x) - (y) - (x - y) = 0,$$
$$(x)^2 + (y)^2 - (x^2 + y^2) = 0,$$
$$(x + y)^2 + (x - y)^2 - 2(x^2 + y^2) = 0,$$
$$4(x)^4 + 4(y)^4 + (x + y)^4 + (x - y)^4 - 6(x^2 + y^2)^2 = 0,$$
$$8(x)^6 + 8(y)^6 + (x + y)^6 + (x - y)^6 - 10(x^2 + y^2)^3 = 0.$$

(5.31)

correspond to six linearly independent abelian relations for the 5-web $\mathcal{W} = \mathcal{W}(u_1, \ldots, u_5)$ on $\mathbb{C}^2$ defined by the following polynomial first integrals:

$$u_1 = x_1, \quad u_2 = y, \quad u_3 = x + y, \quad u_4 = x - y \quad \text{and} \quad u_5 = x^2 + y^2$$

(the abelian relation associated with the first identity of (5.31) is $(du_1, du_2, -du_3, 0, 0)$, etc; see also Example 2.2.7).

It follows that $\mathcal{W}$ is of maximal rank 6 and that the following maps

$$Z_1(t) = \left(1, 1, 2t, 0, 16t^3, 48t^5\right)$$

$$Z_2(t) = \left(1, -1, 2t, 0, 16t^3, 48t^5\right)$$

$$Z_3(t) = \left(-1, 0, 0, 2t, 4t^3, 6t^5\right)$$

$$Z_4(t) = \left(0, -1, 0, 2t, 4t^3, 6t^5\right)$$

$$\text{and } Z_5(t) = \left(0, 0, -1, -2, -12t, -30t^2\right)$$

are such that $Z_i \circ u_i$ is a lift of the ith canonical map κ_i of $\mathcal{W}$, for every $i \in \underline{5}$. The following rational map is then a lift of the associated Poincaré–Blaschke's map $PB_{\mathcal{W}}$ to $\mathbb{C}^\pi$:

$$\tilde{Z} : \mathbb{C}^2 \times \mathbb{P}^1 \dashrightarrow \mathbb{C}^\pi$$

$$((x, y), t) \longmapsto \sum_{i=1}^{6} \frac{Z_i\big(u_i(x, y)\big)}{\partial_x(u_i)\, t - \partial_y(u_i)}.$$

By an elementary computation, one verifies that generically, one has

$$\dim\left(\tilde{Z}, \frac{\partial \tilde{Z}}{\partial x}, \frac{\partial \tilde{Z}}{\partial y}, \frac{\partial \tilde{Z}}{\partial t}\right) = 4.$$

This implies that $PB_{\mathcal{W}}$ has rank 3 at a generic point. Hence the planar 5-web $\mathcal{W}$ is not algebraizable although it is of maximal rank.

Chapter 6
Exceptional Webs

This last chapter is devoted to planar webs of maximal rank. More specifically, it surveys the current state of the art concerning exceptional planar webs.

In Sect. 6.1 a criterion to decide whether or not a given planar web is linearizable is presented. The criterion is phrased in terms of a projective connection adapted to the web under study. The lack of conciseness of the presentation carried out here is balanced by the geometric intuition it may help to build. As corollaries, the existence of exceptional webs and an algebraization result for planar webs, are obtained in Sects. 6.1.4 and 6.1.5, respectively.

Section 6.2 deals with planar webs with infinitesimal automorphisms. Using the result previously laid down in Sect. 2.2.2 of Chap. 2 on the structure of the space of abelian relations of a web $\mathcal{W}$ carrying a transverse infinitesimal automorphism X, the rank of the web obtained from the superposition of $\mathcal{W}$ and the foliation $\mathcal{F}_X$ determined by X, is computed as a function of the rank of $\mathcal{W}$. As a corollary, the existence of exceptional k-webs for arbitrary $k \geq 5$ is settled. Some place is taken to recall some basic definitions from differential algebra, and make precise a problem on the nature of abelian relations that the authors think has some interest.

Section 6.3 starts with basic facts from Cartan–Goldsmith–Spencer theory of linear differential systems. This theory is applied to a differential system having abelian relations of a given planar web as solutions, in order to obtain a computational criterion which decides whether or not the web under study has maximal rank. The approach followed there is a mix of the classical one by Pantazi with the more recent by Hénaut. The necessary criterion for the maximality of the rank by Mihăileanu is briefly discussed without proof.

In Sect. 6.4 some recent classification results obtained by the authors are stated, and the proof of one of them is outlined.

Finally in Sect. 6.5 all the exceptional webs known up to this day, to the best of the authors knowledge, are collected.

© Springer International Publishing Switzerland 2015
J.V. Pereira, L. Pirio, *An Invitation to Web Geometry*, IMPA Monographs 2,
DOI 10.1007/978-3-319-14562-4_6

6.1   Criterion for Linearizability

Throughout this section $\mathcal{W} = \mathcal{W}(\omega_1, \ldots, \omega_k)$ is a germ of smooth k-web on $(\mathbb{C}^2, 0)$. For $i \in \underline{k}$, let $X_i \in T(\mathbb{C}^2, 0)$ be a germ of smooth vector field defining the same foliation as ω_i. Thus $\omega_i(X_i) = 0$ and $X_i(0) \neq 0$.

6.1.1   *Characterization of Linear Webs*

Recall that the web $\mathcal{W}$ is **linear** if all its leaves are contained in affine lines of $\mathbb{C}^2$. Recall also that $\mathcal{W}$ is **linearizable** if it is equivalent to a linear web.

It is rather simple to characterize the linear webs. It suffices to notice that a curve $C \subset (\mathbb{C}^2, 0)$ is contained in a line if and only if for any parametrization $\gamma : (\mathbb{C}, 0) \to C$ the following identity holds true

$$\gamma' \wedge \gamma'' \equiv 0.$$

Therefore, every orbit of a vector field X will be linear, if and only if the determinant

$$\det \begin{pmatrix} X(x) & X(y) \\ X^2(x) & X^2(y) \end{pmatrix}$$

vanishes identically (here $X^2(f)$ stands for $X(X(f))$ for any $f \in \mathcal{O}(\mathbb{C}^2, 0)$).

Indeed, if $\gamma(t) = (\gamma_1(t), \gamma_2(t))$ satisfies $X(\gamma)(t) = \gamma'(t)$ for any $t \in (\mathbb{C}, 0)$, then

$$\det \begin{pmatrix} X(x) & X(y) \\ X^2(x) & X^2(y) \end{pmatrix} (\gamma(t)) = \det \begin{pmatrix} \gamma_1'(t) & \gamma_2'(t) \\ \gamma_1''(t) & \gamma_2''(t) \end{pmatrix}.$$

Therefore the considered web $\mathcal{W}$ is linear if and only if

$$\det \begin{pmatrix} X_i(x) & X_i(y) \\ X_i^2(x) & X_i^2(y) \end{pmatrix} = 0$$

for every $i \in \underline{k}$.

§

To characterize linearizable webs one is naturally led to less elementary considerations. Below, the approach laid down in Sect. 27 of Blaschke–Bol's book [17] is presented in a modern language. For more recent references, see [4, 72, 110].

The inherent difficulty in obtaining an analytic criterion characterizing linearizable webs arises from the fact that the usual method to treat linearization problems comes from differential geometry and is well adapted to deal with continuous families of foliations on $(\mathbb{C}^2, 0)$. But a web is not built from a continuous family of foliations but by a finite number of them. It is the contrast between the finiteness and the continuity that makes the linearization of webs a non-trivial question. For instance, despite many efforts, Gronwall's conjecture is still unsettled:

Conjecture 1 (Gronwall's Conjecture). *If $\mathcal{W}$ and $\mathcal{W}'$ are germs of linear 3-webs on $(\mathbb{C}^2, 0)$ with non-vanishing curvature and if $\varphi : (\mathbb{C}^2, 0) \to (\mathbb{C}^2, 0)$ is a germ of biholomorphism sending $\mathcal{W}$ to $\mathcal{W}'$, then φ is the germification of a projective automorphism of $\mathbb{P}^2$. In other words, a non-hexagonal 3-web admits at most one linearization.*

The interested reader can consult [60, 65, 89, 132] for recent advances on this very simple, quite old but still open conjecture.

In sharp contrast, the equivalent statement for planar k-webs with $k \geq 4$ is true as will be explained below. The point is that, when $k = 4$, it is possible to interpolate the defining foliations by a *unique* second order differential equation cubic in the first derivative. When $k = 3$, although it is possible to interpolate in the same way as for $k = 4$, the lack of uniqueness adds an additional layer of difficulty to the problem.

6.1.2 Affine and Projective Connections

A germ of holomorphic **affine connection** on $(\mathbb{C}^2, 0)$ is a map

$$\nabla : T(\mathbb{C}^2, 0) \to \Omega^1(\mathbb{C}^2, 0) \otimes T(\mathbb{C}^2, 0)$$

satisfying

1. $\nabla(\zeta + \zeta') = \nabla(\zeta) + \nabla(\zeta')$;
2. $\nabla(f\zeta) = f\nabla(\zeta) + df \otimes \zeta$;

for every germs of vector fields $\zeta, \zeta' \in T(\mathbb{C}^2, 0)$ and every germ of function $f \in \mathcal{O}(\mathbb{C}^2, 0)$.

Beware that the map ∇ is not $\mathcal{O}(\mathbb{C}^2, 0)$-linear. In particular, the image of a vector field ζ is a tensor which, at a given point $p \in (\mathbb{C}^2, 0)$, is determined by the whole germ of ζ at p and not only by its value at p.

If $\chi = (\chi_1, \chi_2)$ is a holomorphic frame on $(\mathbb{C}^2, 1)$ and if $\omega = (\omega^1, \omega^2)$ stands for the dual coframe, then ∇ is completely determined by its **Christoffel's symbols** Γ_{ij}^k (relative to the frame χ), defined by the relations

$$\nabla(\chi_j) = \sum_{i,k=1}^{2} \Gamma_{ij}^k \, \omega^i \otimes \chi_k \qquad \text{for } j = 1, 2.$$

Although ∇ when evaluated at a vector field ζ does depend on the germ of ζ as explained above, if $C \subset (\mathbb{C}^2, 0)$ is a curve and ζ is a vector field tangent to it, then the pull-back of $\nabla(\zeta)$ on C is completely determined by the restriction of ζ to C. More precisely, $\nabla(\zeta)$ naturally determines a section of $\Omega^1(C) \otimes T(\mathbb{C}^2, 0)$ which only depends on the restriction of ζ to C. Indeed, if ζ, ζ' and ζ'' are vector fields such that $\zeta - \zeta' = f \zeta''$ where $f \in \mathcal{O}(\mathbb{C}^2, 0)$ is a defining equation for C, then

$$\nabla(\zeta) - \nabla(\zeta') = f \nabla(\zeta'') + df \otimes \zeta''$$

which clearly vanishes on C when contracted with vector fields tangent to C.

A smooth curve $C \subset (\mathbb{C}^2, 0)$ is a **geodesic** of ∇, if for every germ of vector field $\zeta \in TC \subset T(\mathbb{C}^2, 0)$, the vector field $\nabla_\zeta(\zeta) := \langle \nabla(\zeta), \zeta \rangle$ still belongs to TC. To wit, if $\gamma : (\mathbb{C}, 0) \to (\mathbb{C}^2, 0)$ is a parametrization of C, then C is a geodesic for ∇ if and only if there exists a holomorphic 1-form $\eta \in \Omega^1(\mathbb{C}, 0)$ such that $\nabla(\gamma') = \eta \otimes \gamma'$. If $\nabla(\gamma')$ vanishes identically on $(\mathbb{C}, 0)$, then γ is called a **geodesic parametrization** of C. We let the reader verify that a geodesic always admits a geodesic parametrization.

Two affine connections on $(\mathbb{C}^2, 0)$ are **projectively equivalent** if they have exactly the same curves as geodesics. This defines an equivalence relation on the space of affine connections on $(\mathbb{C}^2, 0)$. By definition, a **projective connection** is an equivalence class of this equivalence relation. The class of an affine connection ∇ will be denoted by $[\nabla]$. A smooth curve $C \subset (\mathbb{C}^2, 0)$ is a **geodesic** of $\Pi = [\nabla]$ if it is a geodesic of the affine connection ∇. Of course, this definition does not depend on the representative ∇ of Π.

Two projective connections are **equivalent** if there exists a germ of biholomorphism φ sending the geodesics of one into the geodesics of the other. The **trivial projective connection** Π_0 is the global projective connection on $\mathbb{P}^2$ having the projective lines as geodesics. A projective connection Π is **flat** (or **integrable**) if it is equivalent to Π_0.

Projective Connections and Ordinary Differential Equations

Lemma 6.1.1. *If Π is a projective connection on $(\mathbb{C}^2, 0)$, then there exists a unique affine connection ∇ on $(\mathbb{C}^2, 0)$ such that*

1. *the affine connection ∇ is a representative of Π, that is $[\nabla] = \Pi$;*
2. *the Christoffel's symbols Γ_{ij}^k ($i, j, k = 1, 2$) of ∇ relative to the standard coframe (dx, dy) on $\mathbb{C}^2$ verify the relations*

$$\Gamma_{ij}^k = \Gamma_{ji}^k \qquad and \qquad \Gamma_{1j}^1 + \Gamma_{2j}^2 = 0 \tag{6.1}$$

for every $i, j, k = 1, 2$.

Proof. Exercise left to the reader. $\square$

From now on, a projective connection Π, as well as its normalized representative ∇ provided by the lemma above, will be fixed.

Let $\gamma : (\mathbb{C}, 0) \to (\mathbb{C}^2, 0)$, $t \mapsto \gamma(t) = (\gamma_1(t), \gamma_2(t))$ be a parametrization of a curve $C \subset (\mathbb{C}^2, 0)$. It is a simple exercise to show the equivalence of the three assertions below:

1. C is a geodesic of Π;
2. C is a geodesic of ∇;
3. there exists a function φ such that

$$\frac{d^2 \gamma_k}{dt^2} + \sum_{i,j=1}^{2} \Gamma_{ij}^k \left(\frac{d\gamma_i}{dt} \right) \left(\frac{d\gamma_j}{dt} \right) = \varphi \frac{d\gamma_k}{dt} \tag{6.2}$$

for $k = 1, 2$.

Under the additional hypothesis that C is transverse to the vertical foliation $\{x = cst.\}$, the inequality $\frac{d\gamma_1}{dt}(0) \neq 0$ holds true and, modulo a change of variables, one can assume that $\gamma_1(t) = t$ for every $t \in (\mathbb{C}, 0)$. It is then a simple matter to eliminate the function φ in (6.2) and deduce the following lemma which can be traced back to Beltrami [13].

Lemma 6.1.2. *The geodesics of Π transverse to $\{x = cte.\}$ can be identified with the solutions of the second order differential equation*

$$(\mathscr{E}_\Pi) \qquad \frac{d^2 y}{dx^2} = A \left(\frac{dy}{dx} \right)^3 + B \left(\frac{dy}{dx} \right)^2 + C \frac{dy}{dx} + D \tag{6.3}$$

where A, B, C, D are expressed in function of the Christoffel's symbols Γ_{ij}^k of ∇ as follows:

$$A = \Gamma_{22}^1, \quad B = 2\Gamma_{12}^1 - \Gamma_{22}^2, \quad C = \Gamma_{11}^1 - 2\Gamma_{12}^2, \quad D = -\Gamma_{11}^2. \tag{6.4}$$

Combining Eqs. (6.1) and (6.4), it follows that the Christoffell's symbols Γ_{ij}^k of the normalized affine connection ∇ can be expressed in terms of the functions A, B, C, D. Taking into account Lemma 6.1.1, one deduces the following proposition.

Proposition 6.1.3. *Once a coordinate system x, y is fixed, a projective connection can be identified with a second order differential equation of the form (6.3).*

It is evident, in no matter which affine coordinate system x, y on $\mathbb{C}^2$, that the second order differential equation $(\mathscr{E}_{\Pi_0})$ associated with the trivial projective connection Π_0 is nothing more than $d^2 y/dx^2 = 0$. Therefore, a projective connection is integrable if and only if the second order differential equation $(\mathscr{E}_\Pi)$ is transformed into the *trivial* one, $d^2 v/du^2 = 0$, through a '*point transformation*' $(x, y) \mapsto (u(x, y), v(x, y))$ (cf. [99] for the terminology used here).

The characterization of the second order differential equations equivalent to the trivial one stated below is due to Liouville [84] and Tresse [131] (see Theorem 12.19 in [99] for a recent reference).

Theorem 6.1.4. *The second order differential equation* $y'' = f(x, y, y')$ *is equivalent to the trivial equation* $d^2y/dx^2 = 0$ *through a point transformation if and only if the function* $F = f(x, y, p)$ *verifies* $\frac{\partial^4 F}{\partial p^4} = 0$ *and*

$$D^2(F_{pp}) - 4D(F_{yp}) - 3F_y F_{pp} + 6F_{yy} + F_p\left(4F_{yp} - D(F_{pp})\right) = 0$$

where $D = \frac{\partial}{\partial x} + p\frac{\partial}{\partial y} + F\frac{\partial}{\partial p}$.

Once one restricts to equations of the form (6.3), which are exactly the ones satisfying the first condition $\frac{\partial^4 F}{\partial p^4} = 0$ of the above theorem, the second condition can be rephrased in more explicit terms.

If one sets

$$
\begin{aligned}
L_1 = {} & 2\,C_{xy} - B_{xx} - 3\,D_{yy} - 6\,DA_x - 3\,AD_x \\
& + 3\,DB_y + 3\,DB_y + CB_x - 2\,CC_y
\end{aligned}
$$

$$
\text{and} \quad
\begin{aligned}
L_2 = {} & 2\,B_{xy} - C_{yy} - 3\,A_{xx} + 6\,AD_y + 3\,DA_y \\
& - 3\,AC_x - 3\,CA_y - BC_y + 2\,BB_x
\end{aligned}
$$

then, according to [22], Liouville has shown that the tensor

$$L = \left(L_1 dx + L_2 dy\right) \otimes (dx \wedge dy)$$

is invariant under point transformations and that the differential equation (6.3) is equivalent to the trivial one if and only if L is identically zero.

6.1.3 Linearization of Planar Webs

Let $\mathcal{W}$ be a smooth k-web on $(\mathbb{C}^2, 0)$. It is **compatible with the projective connection** Π if the leaves of $\mathcal{W}$ are geodesics of Π. This is clearly a geometric property: if φ is a biholomorphism, then $\mathcal{W}$ is compatible with $\Pi = [\nabla]$ if and only if $\varphi^*\mathcal{W}$ is compatible with $\varphi^*\Pi = [\varphi^*\nabla]$.

Lemma 6.1.5. *A web* $\mathcal{W}$ *is linear if and only if it is compatible with the trivial projective connection* Π_0. *Consequently,* $\mathcal{W}$ *is linearizable if and only if it is compatible with a flat projective connection.*

It is on this tautology that the linearization criterion presented below is based. It seems appropriate to mention here that this elementary fact is again the core of the solution to the linearizability problem for smooth webs of hypersurfaces in arbitrary dimension, see [110].

Proposition 6.1.6. *If $\mathcal{W}$ is a smooth 4-web on $(\mathbb{C}^2, 0)$, then*

(a) there is a unique projective connection $\Pi_{\mathcal{W}}$ compatible with $\mathcal{W}$;
(b) the web $\mathcal{W}$ is linearizable if and only if $\Pi_{\mathcal{W}}$ is flat.

Proof. It is clear that (b) follows from (a) combined with Lemma 6.1.5.

To prove (a), it is harmless to assume that the leaves of $\mathcal{W}$ are transverse to the line $\{x = 0\}$. The foliation defining $\mathcal{W}$ is thus defined by vector fields X_i (for $i = 1, 2, 3, 4$) which can be written as $X_i = \frac{\partial}{\partial x} + e_i \frac{\partial}{\partial y}$ with $e_i \in \mathcal{O}(\mathbb{C}^2, 0)$.

The orbits of the vector fields X_i are solutions of the second order differential equation

$$y'' = A(y')^3 + B(y')^2 + Cy' + D \tag{6.5}$$

if and only if

$$X_i(e_i) = A(e_i)^3 + B(e_i)^2 + C\, e_i + D \tag{6.6}$$

holds true on $(\mathbb{C}^2, 0)$ for $i = 1, \ldots, 4$.

Consider (6.6) as a system of linear equations in the variables A, B, C, D. The determinant of the associated homogeneous linear system is the Vandermonde determinant

$$\det \begin{bmatrix} 1 & e_1 & (e_1)^2 & (e_1)^3 \\ 1 & e_2 & (e_2)^2 & (e_2)^3 \\ 1 & e_3 & (e_3)^2 & (e_3)^3 \\ 1 & e_4 & (e_4)^2 & (e_4)^3 \end{bmatrix} = \prod_{1 \le i < j \le 4} (e_j - e_i).$$

Since $\mathcal{W}$ is smooth, $e_i(0) \ne e_j(0)$ whenever $i \ne j$. Therefore the Vandermonde determinant above is non-zero and consequently there exists a unique second order differential equation of the form (6.5) admitting the orbits of X_i for $i = 1, \ldots, 4$, as solutions. Item (a) follows from Proposition 6.1.3. □

Corollary 6.1.7. *If $\mathcal{W}$ is a smooth k-web $\mathcal{W}$ on $(\mathbb{C}^2, 0)$ with $k \ge 4$, then the following assertions are equivalent:*

1. $\mathcal{W}$ is linearizable;
2. there exists a flat projective connection Π such that $\Pi = \Pi_{\mathcal{W}'}$ for every 4-subweb $\mathcal{W}'$ of $\mathcal{W}$.

Remark 6.1.8. If $X_i = \frac{\partial}{\partial x} + e_i \frac{\partial}{\partial y}$ (for $i \in \underline{k}$) are vector fields defining a smooth k-web $\mathcal{W}$ then, mimicking the proof of Proposition 6.1.6, it is possible to prove the existence of a unique differential equation of the form $y'' = F(x, y, y')$, polynomial in y', satisfying the following conditions:

1. the p-degree of $F(x, y, p)$ is at most $k - 1$; and
2. the leaves of $\mathcal{W}$ are solutions of $y'' = F(x, y, y')$.

The previous corollary can be rephrased as follows: $\mathcal{W}$ is linearizable if and only if the degree of F is at most three, and if its coefficients satisfy the conditions $L_1 = L_2 = 0$ of Theorem 6.1.4. Since the determination of the differential equation is purely algebraic and can be carried over rather easily, this provides an effective computational test to decide whether or not a given web is linearizable (cf. [72]). The drawback is that, in contrast with the case when $k = 4$, the differential equation obtained when $k > 4$ does not behave nicely under arbitrary change of coordinates $(x, y) \mapsto (u(x, y), v(x, y))$ as a simple computation shows. Indeed, it can be verified that in the resulting equation $v'' = G(u, v, v')$ the function $G(u, v, q)$ may no longer be polynomial, but only rational, in the variable q.

Corollary 6.1.9. *Let $\mathcal{W}$ be a smooth linearizable k-web on $(\mathbb{C}^2, 0)$ with $k \geq 4$. Modulo projective transformations, it admits a unique linearization: if φ, ψ are germs of biholomorphisms such that $\varphi^* \mathcal{W}$ and $\psi^* \mathcal{W}$ are linear webs, then there exists a projective transformation $g \in PGL_3(\mathbb{C})$ for which $\psi = g \circ \varphi$.*

Proof. The hypothesis implies that $\mu = \varphi \circ \psi^{-1}$ verifies $\mu^* \Pi_0 = \Pi_0$. In other words, if $U \subset \mathbb{P}^2$ is an open subset where μ is defined then for every line $\ell \subset \mathbb{P}^2$ intersecting U, there exists a line ℓ_μ such that $\mu(U \cap \ell) = \mu(U) \cap \ell_\mu$. The fundamental theorem of projective geometry implies that μ is the restriction on U of a projective transformation. $\square$

6.1.4 The First Example of an Exceptional Web

Corollary 6.1.9 is rather useful to prove that a given web is not linearizable. Most of the examples of exceptional webs known to this day are the superposition of an algebraic, in particular linear, k-web with one or more non-linear foliations. For example, Bol's 5-web is the superposition of the algebraic 4-web dual to four lines in general position and of a non-linear foliation: the pencil of conics through the four dual points. It follows from Corollary 6.1.9 that $\mathcal{B}_5$ is non-algebraizable.

On the other hand, Bol realized that the rank of $\mathcal{B}_5$ is six. If $\mathcal{B}_5$ is presented as the web

$$\mathcal{B}_5 = \mathcal{W}\left(x, y, \frac{x}{y}, \frac{1-y}{1-x}, \frac{x(1-y)}{y(1-x)}\right)$$

then the abelian relations coming from its 3-subwebs generate a subspace of $\mathcal{A}(\mathcal{B}_5)$ of dimension 5. Moreover, Bol found one extra abelian relation, which can be

written in integral form using the logarithm and Euler's dilogarithm,[1] which is
essentially equivalent to Abel's functional equation for the dilogarithm. Explicitly,

$$R(x) - R(y) - R\left(\frac{x}{y}\right) - R\left(\frac{1-y}{1-x}\right) + R\left(\frac{x(1-y)}{y(1-x)}\right) = 0$$

for every reals x, y such that $0 < x < y < 1$, where R stands for **Rogers
dilogarithm** defined by $R(x) = \mathbf{Li}_2(x) + \frac{1}{2}\log(x)\log(1-x) - \pi^2/6$ for $x \in]0, 1[$.

Although $\mathcal{B}_5$ was the first exceptional web to appear in the literature, there are
simpler examples. For instance, the 5-webs presented in Example 2.2.4 of Chap. 2
all are the superposition of four pencils of lines and one non-linear foliation, thus
are non-linearizable. Since they all have rank six, they are exceptional webs.

6.1.5 Algebraization of Planar Webs

Let now $\mathcal{W} = \mathcal{F}_1 \boxtimes \cdots \boxtimes \mathcal{F}_k$ be a smooth k-web on $(\mathbb{C}^2, 0)$. In contrast with
the higher dimensional case, no hypothesis on $\mathcal{A}(\mathcal{W})$ is needed to assume that the
foliation $\mathcal{F}_i$ is induced by $\omega_i = \varpi^0 + \theta_i \varpi^1$, where $\varpi = (\varpi^0, \varpi^1)$ is a coframe
and $\theta_1, \ldots, \theta_k$ are functions on $(\mathbb{C}^2, 0)$ (with no loss of generality, one can even
assume that $\varpi^0 = dx$ and $\varpi^1 = dy$).

Proposition 6.1.10. *The assertions below are equivalent:*

1. *$\mathcal{W}$ is compatible with a projective connection Π;*
2. *there exist functions N_0, N_1, N_2, N_3 such that*

$$\{d\theta_i\}^1 - \theta_i\{d\theta_i\}^0 = \sum_{\rho=0}^{3}(\theta_i)^\rho N_\rho \qquad (6.7)$$

for every $i \in \underline{k}$.

Proof. Let ∇ be an affine connection representing a projective connection Π, and
let Γ_{ij}^k (with $i, j, k = 0, 1$) be its Christoffel's symbols relative to the coframe $\chi = (\chi_0, \chi_1)$ dual to the coframe ϖ. For $i \in \underline{k}$, set $X_i = \theta_i \chi_0 - \chi_1$: it is a section of $T\mathcal{F}_i$
vanishing nowhere. If $i \in \underline{k}$ is fixed, then the leaves of $\mathcal{F}_i$ are geodesics of Π if and
only if there exists $\zeta_i \in \Omega^1(\mathbb{C}^2, 0)$ satisfying $\nabla(X_i) = \zeta_i \otimes X_i$. More explicitly

$$d\theta_i \otimes \chi_0 + \theta_i \nabla(\chi_0) - \nabla(\chi_1) = \zeta_i \otimes \left(\theta_i \chi_0 - \chi_1\right).$$

---

[1] Euler's dilogarithm is the function $\mathbf{Li}_2(z) = \sum_{n=0}^{\infty} z^n/n^2$. The series converges for $|z| < 1$ and
has analytic continuations along all paths contained in $\mathbb{C} \setminus \{0, 1\}$.

After decomposing this relation in the basis $\chi_p \otimes \omega^q$ (with $p, q = 0, 1$) and using the notation $\zeta_i = \{\zeta_i\}^0 \varpi^0 + \{\zeta_i\}^1 \varpi^1$ for every $i \in \underline{k}$, it is a simple matter to deduce the following four scalar equations:

$$\{d\theta_i\}^0 + \theta_i \Gamma_{00}^0 - \Gamma_{01}^0 = \theta_i \{\zeta_i\}^0$$

$$\{d\theta_i\}^1 + \theta_i \Gamma_{10}^0 - \Gamma_{11}^0 = \theta_i \{\zeta_i\}^1$$

$$\theta_i \Gamma_{00}^1 - \Gamma_{01}^1 = -\{\zeta_i\}^0$$

$$\theta_i \Gamma_{10}^1 - \Gamma_{11}^1 = -\{\zeta_i\}^1. \qquad (6.8)$$

Notice that the last two equations determine $\{\zeta_i\}^0$ and $\{\zeta_i\}^1$. Plugging them into the first two equations allows us to deduce the following: if the leaves of $\mathcal{F}_i$ are geodesics for Π, then θ_i verifies

$$\{d\theta_i\}^1 - \theta_i \{d\theta_i\}^0 = A + \theta_i B + (\theta_i)^2 C + (\theta_i)^3 D \qquad (6.9)$$

where

$$A = \Gamma_{11}^0 \qquad\qquad C = \Gamma_{00}^0 - \Gamma_{10}^1 - \Gamma_{01}^1 \qquad (6.10)$$

$$B = \Gamma_{11}^1 - \Gamma_{10}^0 - \Gamma_{01}^0 \qquad D = \Gamma_{00}^1.$$

Hence the first assertion does imply the second.

Conversely, if the second assertion holds true—which is clearly equivalent to the validity of (6.9)—let ∇ be the affine connection with Christoffel's symbols (in the coframe ϖ) determined by (6.10) and the system of four equations (6.4). It is a simple matter to verify that the result is indeed an affine connection which represents a projective connection having the leaves of the web as geodesics. $\qquad\square$

The next result, when $k = 5$, is discussed in Sect. 30 of the book [17]. In its most general form, it has been obtained by Hénaut in [73]. He phrased it in a slightly different form and under the stronger assumption that the rank of $\mathcal{W}$ is maximal, but his proof still works under the weaker assumption stated below.

Theorem 6.1.11. *Let $\mathcal{W}$ be a smooth k-web on $(\mathbb{C}^2, 0)$ with $k \geq 4$. If*

1. $\dim \mathcal{A}(\mathcal{W})/F^2\mathcal{A}(\mathcal{W}) = 2k - 5$; and
2. $\mathcal{W}$ is compatible with a projective connection

then $\mathcal{W}$ is algebraizable.

Proof. If $k \geq 4$ and $\dim \mathcal{A}(\mathcal{W})/F^2\mathcal{A}(\mathcal{W}) = 2k - 5$, then it is possible to construct a Poincaré-Blaschke map $PB_{\mathcal{W}} : (\mathbb{C}^2, 0) \times \mathbb{P}^1 \rightarrow \mathbb{P}^{2k-6}$. If the rank of this map is two, then $\mathcal{W}$ is algebraizable, according to Theorem 5.5.1.

The key result to establish that $PB_{\mathcal{W}}$ has rank 2 is Lemma 5.3.12. Its proof is based on the relations (5.10) and (5.11) of Lemma 5.3.7. Recall that relations (5.10) do hold true, no matter in which dimension. A careful reading of the proof of

Lemma 5.3.12 reveals that to prove it, one just needs to have relations similar to (5.11) but with the summation in the right hand-side being allowed to range from 0 to $n+1$ instead of from 0 to n (this follows from the fact that $L = n+1$ is allowed in (5.22)).

But, according to Proposition 6.1.10, the existence of such relations is equivalent to the compatibility of $\mathcal{W}$ with a projective connection. Thus the proof of Trépreau's Theorem presented in the previous chapter works as well in dimension two when $\mathcal{W}$ is compatible with a projective connection. The theorem follows.   □

6.2 Infinitesimal Automorphisms

During almost 70 years, Bol's web $\mathcal{B}_5$ has been the unique known example of a non-algebraic web of maximal rank. It enjoys many other properties beside this one. For instance, it admits the symmetric group $\mathfrak{S}_5$ as a group of birational symmetries, the latter acting transitively on the set of its foliations.

If one believes that an exceptional web is a very particular object, one can expect that such a web must enjoy other remarkable properties beside being of maximal rank. It is therefore natural to look for webs of maximal rank admitting non-trivial symmetries. In [91] planar webs of maximal rank admitting a continuous family of automorphisms are considered. The main result obtained in [91] is the explicit construction of continuous families of exceptional k-webs, for any $k \geq 5$. The present section is essentially devoted to these webs.

In what follows, we first study the abelian relations of planar webs admitting an infinitesimal automorphism. For these webs, it is possible to answer precisely to a natural question concerning the nature of the abelian relations. Then, in Sect. 6.2.2, we study the rank of such webs. Finally in Sect. 6.2.3, we construct families of exceptional webs of this kind.

6.2.1 Nature of the Abelian Relations of a Web

As already mentioned, exceptional planar webs and their abelian relations are still mysterious to-date. For instance, even for a web defined by rational first integrals it is not known what kind of transcendency its abelian relations could have. To formulate this question more precisely it is useful to start with recalling some basic definitions of differential algebra.

Basics on Differential Algebra

Recall that a **differential field**[2] is a pair $(\mathbb{K}, \Delta)$, where $\mathbb{K}$ is a field containing $\mathbb{C}$, and Δ is a finite collection of $\mathbb{C}$-derivations of $\mathbb{K}$ subject to the conditions

(a) any two derivations in Δ commute;
(b) the **field of constants** of Δ, that is the intersection of the kernels of the derivations in Δ, is equal to $\mathbb{C}$.

A **differential extension** of $(\mathbb{K}, \Delta)$ is a differential field $(\mathbb{K}_0, \Delta_0)$ such that $\mathbb{K}_0$ is a field extension of $\mathbb{K}$, and for every $\partial_0 \in \Delta_0$ there exists a unique $\partial \in \Delta$ satisfying $\partial = \partial_0|_{\mathbb{K}}$. A differential extension $(\mathbb{K}_0, \Delta_0)$ of $(\mathbb{K}, \Delta)$ is said to be **primitive** if there exists an element $h \in \mathbb{K}_0$ such that $\mathbb{K}_0 = \mathbb{K}(h)$.

The simplest kind of differential extension is when $\mathbb{K}_0$ is an algebraic field extension of $\mathbb{K}$. Extensions of this kind are called **algebraic extensions**.

Another particularly simple kind of differential extensions are the so-called **Liouvillian extensions**. A differential field $(\mathbb{K}', \Delta')$ is a Liouvillian extension of $(\mathbb{K}, \Delta)$, if there exists a finite sequence of differential extensions

$$(\mathbb{K}, \Delta) = (\mathbb{K}_1, \Delta_1) \subset \cdots \subset (\mathbb{K}_r, \Delta_r) \subset (\mathbb{K}_{r+1}, \Delta_{r+1}) = (\mathbb{K}', \Delta')$$

such that for each $i \in \underline{r}$, $\mathbb{K}_{i+1} = \mathbb{K}_i(h_{i+1})$ for some $h_{i+1} \in \mathbb{K}_{i+1}$ satisfying one of the following conditions:

(a) h_{i+1} is algebraic over $\mathbb{K}_i$, or;
(b) for every $\partial \in \Delta_{i+1}$, ∂h_{i+1} belongs to $\mathbb{K}_i$, or;
(c) for every $\partial \in \Delta_{i+1}$, $\frac{\partial h_{i+1}}{h_{i+1}}$ belongs to $\mathbb{K}_i$.

If $(\mathbb{K}, \Delta) = (\mathbb{C}((x, y)), \{\partial_x, \partial_y\})$ is the differential field of germs of meromorphic functions at the origin of $\mathbb{C}^2$ endowed with the natural derivations ∂_x, ∂_y, then a primitive Liouvillian extension over it is obtained by taking (a) a primitive algebraic extension; or (b) the integral of a closed meromorphic 1-form; or (c) the exponential of the integral of a closed meromorphic 1-form.

If $(\mathbb{K}, \{\partial_x, \partial_y\})$ is a differential subfield of $(\mathbb{C}((x, y)), \{\partial_x, \partial_y\})$ then, by definition, a web $\mathcal{W}$ on $(\mathbb{C}^2, 0)$ is **defined over** $\mathbb{K}$ if there exists a k-symmetric 1-form with coefficients in $\mathbb{K}$ defining $\mathcal{W}$. More explicitly, there exists

$$\omega = \sum_{i+j=k} a_{ij}(x, y) dx^i dy^j \in \mathrm{Sym}^k \Omega^1(\mathbb{C}^2, 0)$$

such that $\mathcal{W} = \mathcal{W}(\omega)$ and $a_{ij} \in \mathbb{K}$ for every pair $(i, j) \in \mathbb{N}^2$ satisfying $i + j = k$. Similarly, an abelian relation of a given smooth web on $(\mathbb{C}^2, 0)$ is defined over $\mathbb{K}$ if

---

[2] Here only differential fields over $\mathbb{C}$ will be considered. Of course, it is possible to deal with more general fields.

its components are 1-forms with coefficients in $\mathbb{K}$. If $\mathcal{W}$ is a web defined over $\mathbb{K}$, then the **field of definition** of its abelian relations is the differential extension of $\mathbb{K}$ generated by all the coefficients of all the components of all the abelian relations of $\mathcal{W}$.

Problem 1. Let $\mathcal{W}$ be a germ of smooth k-web on $(\mathbb{C}^2, 0)$. If $\mathcal{W}$ is defined over $\mathbb{K}$, what can be said about the field of definition of its abelian relations? Is it a Liouvillian extension of $\mathbb{K}$?

If $\mathcal{W}$ is a hexagonal 3-web on $(\mathbb{C}^2, 0)$, then its unique abelian relation is defined over a Liouvillian extension of its field of definition. An argument has already been given in the proof of the implication (b) $\Longrightarrow$ (c) of Theorem 1.2.4.

An Answer, But in a Very Particular Case

Due to the apparent difficulty of Problem 1, one could propose to lower the dimension of the ambient space. At first sight this seems to be pure nonsense since it doesn't appear to be reasonable to talk about webs on $(\mathbb{C}, 0)$. A way out is to interpret less strictly the *lowering* of the dimension. In the study of systems of differential equations, the usual setup where one is allowed to lower dimensions is when the system possesses infinitesimal symmetries. Having this vague discourse in mind, it suffices to look back at Chap. 2 to realize that it has been formally implemented in the proof of Proposition 2.2.5. In particular, one obtains as a corollary the following

Proposition 6.2.1. *Let* $(\mathbb{K}, \{\partial_x, \partial_y\})$ *be a differential subfield of* $(\mathbb{C}((x, y)),$ $\{\partial_x, \partial_y\})$, *and let* $\mathcal{W}$ *be a germ of smooth k-web defined over $\mathbb{K}$. If $\mathcal{W}$ admits a transverse infinitesimal automorphism, also defined over $\mathbb{K}$, then there exists a Liouvillian extension of $\mathbb{K}$ over which all the abelian relations of $\mathcal{W}$ are defined.*

Proof. It follows immediately from Proposition 2.2.5. $\square$

6.2.2 Variation of the Rank

Proposition 2.2.5 also allows to compare the rank of a web $\mathcal{W}$ admitting a transverse infinitesimal automorphism X, with the rank of the web $\mathcal{W} \boxtimes \mathcal{F}_X$ obtained by the superposition of $\mathcal{W}$ and the foliation induced by X, denoted by $\mathcal{F}_X$.

Theorem 6.2.2. *Let $\mathcal{W}$ be a smooth k–web which admits a transverse infinitesimal automorphism X. Then*

$$\mathrm{rank}(\mathcal{W} \boxtimes \mathcal{F}_X) = \mathrm{rank}(\mathcal{W}) + (k - 1).$$

In particular, $\mathcal{W}$ is of maximal rank if and only if $\mathcal{W} \boxtimes \mathcal{F}_X$ is of maximal rank.

Proof. Let $W = W(\omega_1, \ldots, \omega_k) = \mathcal{F}_1 \boxtimes \cdots \boxtimes \mathcal{F}_k$. Recall from Chap. 2 that the canonical first integral of $\mathcal{F}_i$ relative to X is

$$u_i = \int \frac{\omega_i}{\omega_i(X)}.$$

In particular, its differential is $\eta_i = du_i = \frac{\omega_i}{\omega_i(X)}$.

Notice that when j varies from 2 to k, the following identities hold true:

$$i_X(\eta_1 - \eta_j) = 0 \qquad \text{and} \qquad L_X(\eta_1 - \eta_j) = 0.$$

Consequently there exists $g_j \in \mathbb{C}\{t\}$ for which

$$du_1 - du_j - g_j(u_{k+1})du_{k+1} = 0, \tag{6.11}$$

where u_{k+1} is a fixed first integral of $\mathcal{F}_X$.

Clearly these are abelian relations for the web $W \boxtimes \mathcal{F}_X$. If $\mathcal{A}_0(W \boxtimes \mathcal{F}_X)$ stands for the eigenspace of L_X associated with the eigenvalue zero, then the aforementioned abelian relations span a vector subspace of it which will be denoted by $\mathcal{V}$. Notice that $\dim(\mathcal{V}) = k - 1$.

Observe that $\mathcal{V}$ fits into the sequence

$$0 \to \mathcal{V} \xrightarrow{i} \mathcal{A}_0(W \boxtimes \mathcal{F}_X) \xrightarrow{L_X} \mathcal{A}_0(W). \tag{6.12}$$

Notice that this sequence is exact. Indeed,

$$K = \ker\left(L_X : \mathcal{A}_0(W \boxtimes \mathcal{F}_X) \to \mathcal{A}_0(W)\right)$$

is generated by abelian relations of the form $\sum_{i=1}^k c_i du_i + h(u_{k+1})du_{k+1} = 0$, where $c_i \in \mathbb{C}$ and $h \in \mathbb{C}\{t\}$. Since $i_X du_i = 1$ for each $i \in \underline{k}$, it follows that the constants c_i satisfy $\sum_{i=1}^k c_i = 0$. This implies that the abelian relations in the kernel of L_X can be written as linear combinations of abelian relations of the form (6.11). Therefore

$$K = \mathcal{V} \tag{6.13}$$

and consequently $\ker(L_X) \subset \operatorname{Im}(i)$. The exactness of (6.12) follows easily.

From general principles one can deduce that the sequence

$$0 \to \frac{\mathcal{V}}{\mathcal{A}_0(W) \cap \mathcal{V}} \xrightarrow{i} \frac{\mathcal{A}_0(W \boxtimes \mathcal{F}_X)}{\mathcal{A}_0(W)} \xrightarrow{L_X} \frac{\mathcal{A}_0(W)}{L_X \mathcal{A}_0(W)}$$

is also exact. Thus to prove the theorem it suffices to verify that

(a) $\mathcal{V}$ is isomorphic to $\frac{\mathcal{V}}{\mathcal{A}_0(\mathcal{W}) \cap \mathcal{V}} \oplus \frac{\mathcal{A}_0(\mathcal{W})}{L_X \mathcal{A}_0(\mathcal{W})}$;

(b) the morphism $L_X : \mathcal{A}_0(\mathcal{W} \boxtimes \mathcal{F}_X) \to \mathcal{A}_0(\mathcal{W})$ is surjective;

(c) the vector spaces $\frac{\mathcal{A}_0(\mathcal{W} \boxtimes \mathcal{F}_X)}{\mathcal{A}_0(\mathcal{W})}$ and $\frac{\mathcal{A}(\mathcal{W} \boxtimes \mathcal{F}_X)}{\mathcal{A}(\mathcal{W})}$ are isomorphic.

The key to verify (a) is the nilpotency of L_X on $\mathcal{A}_0(\mathcal{W})$. It implies that $\frac{\mathcal{A}_0(\mathcal{W})}{L_X \mathcal{A}_0(\mathcal{W})}$ is isomorphic to $\mathcal{A}_0(\mathcal{W}) \cap K$. Combining this with (6.13) assertion (a) follows.

To prove assertion (b) it suffices to construct a map $\Phi : \mathcal{A}_0(\mathcal{W}) \to \mathcal{A}_0(\mathcal{W} \boxtimes \mathcal{F}_X)$ such that $L_X \circ \Phi = \mathrm{Id}$. Proposition 2.2.5 implies that $\mathcal{A}_0(\mathcal{W})$ is spanned by abelian relations of the form $\sum_{i=1}^{k} c_i (u_i)^r d u_i = 0$, where the $c_1, \ldots, c_k$ are complex numbers and r is a non-negative integer.

Since

$$\sum_{i=1}^{k} c_i (u_i)^r d u_i = \frac{1}{r+1} L_X \left(\sum_{i=1}^{k} c_i (u_i)^{r+1} d u_i \right) = 0,$$

there exists a unique function $H \in \mathbb{C}\{t\}$ satisfying

$$\sum_{i=1}^{k} c_i (u_i)^{r+1} d u_i + H(u_{k+1}) d u_{k+1} = 0.$$

If one sets

$$\Phi \left(\sum_{i=1}^{k} c_i (u_i)^r d u_i \right) = \frac{1}{r+1} \left(\sum_{i=1}^{k} c_i (u_i)^{r+1} d u_i + H(u_{k+1}) d u_{k+1} \right),$$

then $L_X \circ \Phi = \mathrm{Id}$ and assertion (b) follows.

To prove assertion (c), first notice that

$$\mathcal{A}(\mathcal{W} \boxtimes \mathcal{F}_X) = \mathcal{A}_0(\mathcal{W} \boxtimes \mathcal{F}_X) \oplus \mathcal{A}_*(\mathcal{W} \boxtimes \mathcal{F}_X)$$

where $\mathcal{A}_*(\mathcal{W} \boxtimes \mathcal{F}_X)$ is the sum of eigenspaces corresponding to non-zero eigenvalues. Of course, the latter is invariant by L_X. Moreover the following equality holds true

$$L_X \left(\mathcal{A}_*(\mathcal{W} \boxtimes \mathcal{F}_X) \right) = \mathcal{A}_*(\mathcal{W} \boxtimes \mathcal{F}_X).$$

On the other hand, L_X *kills* the component of an abelian relation corresponding to the foliation $\mathcal{F}_X$. In particular

$$L_X \left(\mathcal{A}_*(\mathcal{W} \boxtimes \mathcal{F}_X) \right) \subset \mathcal{A}_*(\mathcal{W}).$$

This is sufficient to show that $\mathcal{A}_*(\mathcal{W} \boxtimes \mathcal{F}_X) = \mathcal{A}_*(\mathcal{W})$ and to deduce assertion (c).

Putting all together, it follows that

$$\text{rank}(\mathcal{W} \boxtimes \mathcal{F}_X) = \text{rank}(\mathcal{W}) + (k - 1).$$

To prove the last claim of the theorem, just remark that the $(k+1)$-web $\mathcal{W} \boxtimes \mathcal{F}_X$ is of maximal rank if and only if

$$\text{rank}(\mathcal{W} \boxtimes \mathcal{F}_X) = \frac{k(k-1)}{2} = \frac{(k-1)(k-2)}{2} + (k - 1).$$

$\square$

Of course, one can also deduce from Theorem 6.2.2 the following analogue of Proposition 6.2.1.

Proposition 6.2.3. *Let* $(\mathbb{K}, \{\partial_x, \partial_y\})$ *be a differential subfield of* $(\mathbb{C}((x, y)), \{\partial_x, \partial_y\})$, *and let* $\mathcal{W}$ *be a germ of smooth k-web defined over* $\mathbb{K}$. *If* $\mathcal{W}$ *admits a transverse infinitesimal automorphism X, also defined over* $\mathbb{K}$, *then there exists a Liouvillian extension of* $\mathbb{K}$ *over which all the abelian relations of* $\mathcal{W} \boxtimes \mathcal{F}_X$ *are defined.*

It should not be very hard to drop the hypothesis on the field of definition of the infinitesimal automorphism of $\mathcal{W}$ in Proposition 6.2.1 as well as in Proposition 6.2.3. More precisely, it could be true that the infinitesimal automorphisms of a web defined over $\mathbb{K}$ should also be defined over $\mathbb{K}$, or at least over a Liouvillian extension of $\mathbb{K}$.

6.2.3 Infinitely Many Families of Exceptional Webs

Let C be a degree k algebraic curve in $\mathbb{P}^2$ invariant by a projective $\mathbb{C}^*$-action $\varphi : \mathbb{C}^* \times \mathbb{P}^2 \to \mathbb{P}^2$. Since this action is assumed to be projective, it comes that φ induces a dual action $\check{\varphi} : \mathbb{C}^* \times \check{\mathbb{P}}^2 \to \check{\mathbb{P}}^2$ which is a one-parameter group of projective automorphisms of the dual k-web $\mathcal{W}_C$. Consequently, the web $\mathcal{W}_C(\ell_0)$, the germification of $\mathcal{W}_C$ at a generic point $\ell_0 \in \check{\mathbb{P}}^2$, admits an infinitesimal automorphism.

It is a simple matter to show that in a suitable projective coordinate system $[x : y : z]$, a plane curve C invariant by a $\mathbb{C}^*$-action is cut out by an equation of the form

$$x^{\epsilon_1} \cdot y^{\epsilon_2} \cdot z^{\epsilon_3} \cdot \prod_{i=1}^{n} (x^a + \lambda_i \, y^b z^{a-b}) = 0 \tag{6.14}$$

where $\epsilon_1, \epsilon_2, \epsilon_3 \in \{0, 1\}$, $n, a, b \in \mathbb{N}$ are such that $n \geq 1$, $a \geq 2$, $1 \leq b \leq a/2$, $\gcd(a, b) = 1$ and the λ_i's are distinct non-zero complex numbers (cf. [6, Section 1] for instance).

For a curve of this form the $\mathbb{C}^*$-action in question is

$$\varphi: \quad \mathbb{C}^* \times \mathbb{P}^2 \longrightarrow \mathbb{P}^2$$
$$\left(t, [x:y:z]\right) \longmapsto \left[t^{b(a-b)}x : t^{a(a-b)}y : t^{ab}z\right].$$

Moreover once $\epsilon_1, \epsilon_2, \epsilon_3, n, a, b$ are fixed, one can always assume that $\lambda_1 = 1$ and in this case the set of $n-1$ complex numbers $\{\lambda_2, \ldots, \lambda_n\}$ projectively characterizes the curve C. In particular, there exists a $(d-1)$-dimensional family of degree $2d$ (or $2d+1$) reduced plane curves which all are projectively distinct and invariant by the same $\mathbb{C}^*$-action: for a given $2d + \delta$ with $\delta \in \{0, 1\}$ set $a = 2$, $b = 1$, $\epsilon_1 = \delta$ and $\epsilon_2 = \epsilon_3 = 0$.

If C is a reduced curve cut out by an equation of the form (6.14), then $\mathcal{W}_C$ is invariant by the $\mathbb{C}^*$-action $\check{\varphi}$ dual to φ. Denote by X the infinitesimal generator of $\check{\varphi}$ and by $\mathcal{F}_X$ the corresponding foliation.

Theorem 6.2.4. *If* $\deg C \geq 4$*, then* $\mathcal{W}_C \boxtimes \mathcal{F}_X$ *is exceptional. Moreover if* C' *is another curve invariant by* φ*, then* $\mathcal{W}_C \boxtimes \mathcal{F}_X$ *is analytically equivalent to* $\mathcal{W}_{C'} \boxtimes \mathcal{F}_X$ *if and only if the curve* C *is projectively equivalent to* C'*.*

Proof. Since $\mathcal{W}_C$ has maximal rank it follows from Theorem 6.2.2 that $\mathcal{W}_C \boxtimes \mathcal{F}_X$ is also of maximal rank. Suppose that its localization at a point $\ell_0 \in \check{\mathbb{P}}^2$ is algebraizable and let $\psi : (\check{\mathbb{P}}^2, \ell_0) \to (\mathbb{C}^2, 0)$ be a holomorphic algebraization. In particular, ψ linearizes $\mathcal{W}_C$.

Since both $\mathcal{W}_C$ and $\psi_*(\mathcal{W}_C)$ are linear k-webs with $k \geq 4$, it follows from Corollary 6.1.9 that ψ is the localization of a projective automorphism. But the generic leaf of $\mathcal{F}_X$ is not contained in any projective line and consequently $\psi_*(\mathcal{W} \boxtimes \mathcal{F}_X)$ is not linear. This concludes the proof of the theorem. □

Remark 6.2.5. It is not known whether the families of examples above are *irreducible components of the space of exceptional webs* in the sense that the generic element does not admit a deformation as an exceptional web that is not of the form $\mathcal{W}_C \boxtimes \mathcal{F}_X$. Due to the presence of infinitesimal automorphisms one could suppose that they are indeed degenerations of some other exceptional webs.

6.3 Pantazi–Hénaut Criterion

If $\mathcal{W}$ is a 3-web on $(\mathbb{C}^2, 0)$, then $\mathcal{W}$ has maximal rank if and only if its curvature $K(\mathcal{W})$ vanishes identically. In this section, a generalization of this result to arbitrary planar webs will be presented. The strategy sketched below can be traced back to Pantazi [101]. Recently, unaware of Pantazi's result, Hénaut [77] proved an essentially equivalent result but formulated in more intrinsic terms.

Even more recently, Cavalier and Lehmann proved that it is possible to extend the Pantazi–Hénaut construction to *ordinary* codimension one webs on $(\mathbb{C}^n, 0)$ for n arbitrary. This text will not deal with this generalization. For details see [27].

Below, after presenting the basics of the theory of linear differential systems, the arguments of [101] are presented using the modern formalism introduced in this context by Hénaut.

6.3.1   Linear Differential Systems

For a more detailed exposition the reader can consult [125] and the references therein.

Jet Spaces

Let E be a rank r vector bundle over $(\mathbb{C}^n, 0)$. Since the setup is local, E is of course trivial. Thus x, p with $x = (x_1, \ldots, x_n) \in (\mathbb{C}^n, 0)$ and $p = (p^1, \ldots, p^r) \in \mathbb{C}^r$ constitute a coordinate system on the total space of E. A section ξ of E will be identified with a map $(\xi^1, \ldots, \xi^r) : (\mathbb{C}^n, 0) \to \mathbb{C}^r$. We will commit the abuse to denote the sheaf of local sections of E by the same notation.

The **space $J^\ell(E)$ of ℓ-jets of sections** of E is the vector bundle over $(\mathbb{C}^n, 0)$ with fiber $J^\ell_x(E)$ over a point $x \in (\mathbb{C}^n, 0)$ equal to the quotient of the space of germs of sections of E at x by the subspace of germs vanishing at x up to order $\ell + 1$. Given a section ξ of E, then its ℓ-jet at x will be denoted by $j^\ell_x(\xi)$. In order to make sense of the following map

$$j^\ell : E \longrightarrow J^\ell(E)$$

one has to think of it not as a map of vector bundles (derivations are not $\mathcal{O}(\mathbb{C}^n, 0)$-linear) but as a morphism of sheaves of $\mathbb{C}$-modules.

On $J^\ell_x(E)$ there is a natural system of linear coordinates:

$$p^s_\sigma(j^\ell_x(\xi)) = \partial_\sigma(\xi^s)(x) = \frac{\partial^{|\sigma|} \xi^s}{\partial x^\sigma}(x)$$

for $s \in \underline{r}$ and $\sigma = (\sigma_a)^n_{a=1} \in \mathbb{N}^n$ with $|\sigma| = \sum_a \sigma_a \le \ell$.

For every ℓ, ℓ' with $\ell \ge \ell'$, there is a natural morphism of vector bundles (unlike j^ℓ, this morphisms is $\mathcal{O}(\mathbb{C}^n, 0)$-linear)

$$\pi^{\ell'}_\ell : J^\ell(E) \longrightarrow J^{\ell'}(E)$$

$$\left(x, j^\ell_x(\xi)\right) \longmapsto \left(x, j^{\ell'}_x(\xi)\right).$$

By convention, $J^q(E)$ is the 0 vector bundle when $q < 0$.

Linear Differential Systems

A **linear differential system of order** q **in** r **indeterminates** is defined as the kernel

$$S = \operatorname{Ker} \Phi \subset J^q(E)$$

of a **linear differential operator of order** q **in** r **indeterminates**, that is a morphism of $\mathcal{O}(\mathbb{C}^n, 0)$-modules

$$\Phi : J^q(E) \longrightarrow F.$$

Explicitly, if $F = (\mathbb{C}^n, 0) \times \mathbb{C}^m$ and $\Phi = (\Phi^1, \dots, \Phi^m) : J^q(E) \longrightarrow F$ with

$$\Phi^\kappa(x, p) = \sum_{s=1}^{r} \sum_{|\sigma| \le q} A_{s,\sigma}^\kappa(x)\, p_\sigma^s$$

for $\kappa \in \underline{m}$, then the sections of $S = \operatorname{Ker} \Phi$ which are also images of sections of E through $j^q : E \to J^q(E)$ correspond to solutions of the system of differential equations

$$(S) \qquad \sum_{s=1}^{r} \sum_{|\sigma| \le q} A_{s,\sigma}^\kappa \frac{\partial^{|\sigma|} \varphi^s}{\partial x^\sigma} = 0 \qquad \kappa \in \underline{m}$$

in the unknowns $\varphi^1, \dots, \varphi^r : (\mathbb{C}^n, 0) \to \mathbb{C}$. Beware that in general $S \subset J^q(E)$ is not a vector subbundle of $J^q(E)$: the rank of the fibers of S may vary from point to point.

If the q-jet of $(\varphi_1, \dots, \varphi_r)$ is in S then its $(q+1)$-jet will be in $S^{(1)} \subset J^{q+1}(E)$, that is the homogeneous linear partial differential equation of order $(q+1)$ deduced from S through derivation with respect to the free variables x_a:

$$(S^{(1)}) \quad \begin{cases} \sum_{s,\,|\sigma| \le q} A_{s,\sigma}^\kappa \dfrac{\partial^{|\sigma|} \varphi^s}{\partial x_\sigma} = 0 & \kappa \in \underline{m}; \\[2ex] \sum_{s,\,|\sigma| \le q} \dfrac{\partial A_{s,\sigma}^\kappa}{\partial x_a} \dfrac{\partial^{|\sigma|} \varphi^s}{\partial x_\sigma} + A_{s,\sigma}^\kappa \dfrac{\partial^{|\sigma|+1} \varphi^s}{\partial x_{\sigma(a)}} = 0 & a \in \underline{n} \end{cases}$$

with $\sigma(a) = (\overline{\sigma}_b)_{b=1}^n$ defined as follows: $\overline{\sigma}_a = \sigma_a + 1$ and $\overline{\sigma}_b = \sigma_b$ if $b \ne a$.

Of course, this operation can be iterated. Setting $S = S^{(0)}$ and

$$S^{(\ell+1)} = \left(S^{(\ell)}\right)^{(1)}$$

for every $\ell \ge 0$, one derives from S a family of linear systems of partial differential equations $S^{(\ell)} \subset J^{q+\ell}(E)$. By definition, $S^{(\ell)}$ is the **ℓ-th prolongation** of S.

It is not hard to construct a morphism of vector bundles

$$\Phi^{(\ell)} : J^{q+\ell}(E) \longrightarrow J^{\ell}(F)$$

such that $S^{(\ell)} = \mathrm{Ker}\,\Phi^{(\ell)}$. The morphism $\Phi^{(\ell)}$ is the ℓ**-th prolongation** of Φ.

Formal Integrability

A differential system S is **regular** if $S^{(\ell)}$ is a vector subbundle of $J^{q+\ell}(E)$ for no matter which $\ell \geq 0$.

Let $\ell \geq 0$ be fixed. The restriction of the natural projection $J^{q+\ell+1}(E) \to J^{q+\ell}(E)$ to $S^{(\ell+1)}$ induces a natural morphism of $\mathcal{O}(\mathbb{C}^n,0)$-modules

$$S^{(\ell+1)} \xrightarrow{\ \overline{\pi}^{q+\ell}_{q+\ell+1}\ } S^{(\ell)}.$$

By definition, S is **formally integrable** if for every $\ell \geq 0$ the morphism $\overline{\pi}^{q+\ell}_{q+\ell+1}$ is surjective. In less precise terms, a system is formally integrable if given a q-jet in S then there exists a $(q + \ell)$-jet in $S^{(\ell)} \subset J^{q+\ell}(E)$ coinciding with the original one up to order q for no matter which $\ell \geq 0$.

A **solution** of S is a section ξ of E such that $j^q(\xi)$ is a section of S. Consequently, $j^{q+\ell}(\sigma)$ is a section of $S^{(\ell)}$ for every $\ell \geq 0$. If $\mathrm{Sol}(S)$ denotes the space of solutions of S, then the surjectivity of $j^q : Sol(S) \to S$ on the fibers is a sufficient condition for the formal integrability of S.

In the analytic category the formal integrability of a differential system is particularly meaningful: Cartan–Kähler theorem ensures the convergence of formal solutions. In the case of linear differential systems, the proof of this result boils down to a simple application of the method of majorants for formal power series.

For $\ell \geq 0$, let $f_1, \ldots, f_\ell \in \mathcal{O}(\mathbb{C}^n,0)^3$ be functions in the maximal ideal $\mathfrak{M}_x$ of x, that is $f_1(x) = \cdots = f_\ell(x) = 0$. Set $f = f_1 \cdots f_\ell$. Since $\mathrm{Sym}^\ell(T^*_x(\mathbb{C}^n,0))$ is generated by the ℓ-symmetric forms $(df_1 \cdots df_\ell)(x)$ as a $\mathbb{C}$-vector space, one can define a linear map

$$\varepsilon^\ell_x : \mathrm{Sym}^\ell\big(T^*_x(\mathbb{C}^n,0)\big) \otimes E \to J^\ell_x(E)$$

through the formula

$$\varepsilon^\ell_x(df_1 \cdots df_\ell \otimes e) = j^\ell(f_1 \cdots f_\ell \cdot e)(x).$$

Because $f \in (\mathfrak{M}_x)^\ell$, $\pi^{\ell-1}_\ell(j^\ell(fe)(x)) = 0$ for every section e of E. Varying x in $(\mathbb{C}^n,0)$, one deduces an injective morphism of vector bundles

---

[3]Recall the convention about germs used throughout. Here $(\mathbb{C}^n,0)$ must be seen as a small open subset containing the origin.

$$\text{Sym}^\ell\big(T^*(\mathbb{C}^n,0)\big) \otimes E \xrightarrow{\ \varepsilon^\ell\ } J^\ell(E)$$

$$\big(df_1 \cdots \cdots df_\ell\big) \otimes e \longmapsto j^\ell\big(f_1 \cdots f_\ell \cdot e\big)$$

which fits into the exact sequence

$$0 \to \text{Sym}^{\ell+1}\big(T^*(\mathbb{C}^n,0)\big) \otimes E \xrightarrow{\ \varepsilon^{\ell+1}\ } J^{\ell+1}(E) \xrightarrow{\ \pi^\ell_{\ell+1}\ } J^\ell(E) \to 0. \qquad (6.15)$$

For $\ell \geq 0$, the ℓ-**th symbol map** of $\Phi : J^q(E) \to F$ is defined as the composition

$$\sigma^{(\ell)}(\Phi) = \Phi^{(\ell)} \circ \varepsilon^{q+\ell} : \text{Sym}^{q+\ell}\big(T^*(\mathbb{C}^n,0)\big) \otimes E \longrightarrow J^\ell(F).$$

The 0-th symbol map is, by convention, $\sigma(\Phi) = \sigma^{(0)}(\Phi)$. By definition,

$$\mathfrak{S}^{(\ell)} = \text{Ker}\,\sigma^{(\ell)}(\Phi) \subset \text{Sym}^{q+\ell}\big(T^*(\mathbb{C}^n,0)\big) \otimes E$$

is the ℓ-**th symbol** of the linear differential system S. It is completely determined by S, that is it does not depend on the presentation of S as the kernel of Φ. From the exact sequence (6.15), it follows that $\mathfrak{S}^{(\ell)}$ fits into the exact sequence of $\mathbb{C}$-sheaves

$$0 \to \mathfrak{S}^{(\ell)} \xrightarrow{\ \varepsilon\ } S^{(\ell)} \xrightarrow{\ \pi\ } S^{(\ell-1)} \qquad (6.16)$$

for every $\ell \geq 0$, with the convention that $S^{(-1)} = J^{q-1}(E)$ and where $\overline{\pi} = \pi^{q+\ell-1}_{q+\ell}\big|_{S^{(\ell)}}$.

Notice that $\mathfrak{S}^{(\ell)}$ is not a vector bundle a priori: it can be naively seen as a family of vector spaces $\{\mathfrak{S}^{(\ell)}(x)\}_{x \in (\mathbb{C}^n,0)}$ but the dimension of $\mathfrak{S}^{(\ell)}(x)$ may vary with x.

It can be verified that $\mathfrak{S}^{(\ell)}$ is completely determined by $\mathfrak{S}$. In particular, if $\mathfrak{S} = 0$, then $\mathfrak{S}^{(\ell)} = 0$ for every $\ell \geq 0$.

Theorem 6.3.1. *Let $S \subset J^q(E)$ be a linear differential system with $\mathfrak{S}_S = 0$. If the natural morphism $S^{(1)} \to S$ is surjective, then S is regular and formally integrable.*

Proof. This is a particular case of a theorem by Goldschmidt, see [125, Theorem 1.5.1]. The proof is omitted. □

Spencer Operator and Connections

The **Spencer operator** (for $s, q \geq 0$)

$$D : \Omega^s \otimes J^q(E) \to \Omega^{s+1} \otimes J^{q-1}(E)$$

is characterized by the following properties:

1. for section ξ of E, one has $D(j^q(\xi)) = 0$;
2. for every $\omega \in \Omega^j$ and every $\eta \in \Omega^* \otimes J^q(E)$, one has

$$D(\omega \wedge \eta) = d\omega \wedge \pi_q^{q-1}(\eta) + (-1)^j \omega \wedge D(\eta).$$

It can be verified that $D \circ D = 0$. Consequently, there is the following complex of $\mathbb{C}$-sheaves

$$0 \to E \xrightarrow{j^q} J^q(E) \xrightarrow{D} \Omega^1 \otimes J^{q-1}(E) \xrightarrow{D} \cdots \xrightarrow{D} \Omega^n \otimes J^{q-n}(E) \to 0 \quad (6.17)$$

from which one can extract the exact sequence

$$0 \to E \xrightarrow{j^q} J^q(E) \xrightarrow{D} \Omega^1 \otimes J^{q-1}(E).$$

To simplify, let us suppose that the dimension n of the basis is 2. If $S \subset J^q(E)$ is a linear differential system over $(\mathbb{C}^2, 0)$, then the restriction of the complex (6.17) to S and its prolongations yields the **first Spencer complex** of $S^{(\ell)}$:

$$0 \to \mathrm{Sol}(S) \xrightarrow{j^{q+\ell}} S^{(\ell)} \xrightarrow{D} \Omega^1 \otimes S^{(\ell-1)} \xrightarrow{D} \Omega^2 \otimes S^{(\ell-2)} \to 0$$

Note that this is not a complex of $\mathcal{O}(\mathbb{C}^2, 0)$-sheaves but only of $\mathbb{C}$-sheaves. It is clarifying to notice the resemblance with the usual de Rham complex $0 \to \mathbb{C} \xrightarrow{d} \Omega^0 \xrightarrow{d} \Omega_1 \xrightarrow{d} \Omega^2 \to 0$.

There is the following commutative diagram

$$
\begin{array}{ccccccccc}
& & & & & & & & 0 \\
& & & & & & & & \downarrow \\
& & & & & & & & \Omega^2 \otimes \mathfrak{S}_S \\
& & & & & & & & \downarrow \\
0 & \longrightarrow & \mathrm{Sol}(S) & \xrightarrow{j^{q+2}} & S^{(2)} & \xrightarrow{D} & \Omega^1 \otimes S^{(1)} & \xrightarrow{D} & \Omega^2 \otimes S \\
& & \| & & \downarrow & & \downarrow & & \downarrow \\
0 & \longrightarrow & \mathrm{Sol}(S) & \xrightarrow{j^{q+1}} & S^{(1)} & \xrightarrow{D} & \Omega^1 \otimes S & \xrightarrow{D} & \Omega^2 \otimes J^{q-1}(E)
\end{array}
$$

with columns and lines being complexes.

Suppose now that $\mathfrak{S}_S^{(1)} = 0$ and that the natural morphism $S^{(1)} \to S$ is surjective. Taken together, these two conditions imply the isomorphism $\overline{\pi} : S^{(1)} \simeq S$. Let us denote again by $\overline{\pi}$ the induced isomorphisms $\Omega^k \otimes S^{(1)} \simeq \Omega^k \otimes S$ for $k = 1, 2$. Setting

$$\nabla = (\overline{\pi})^{-1} \circ D : S^{(1)} \to \Omega^1 \otimes S^{(1)}$$

one defines a linear connexion ∇ on $S^{(1)}$ such that the diagram below commutes.

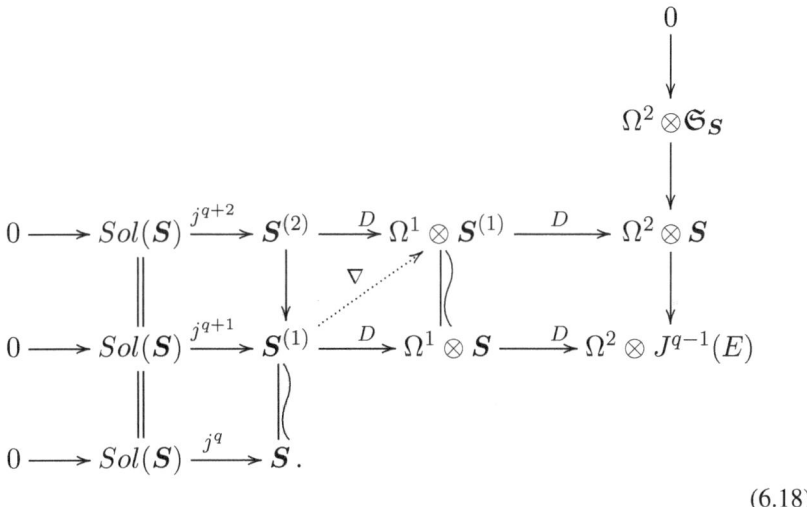

$$(6.18)$$

A simple verification shows that the curvature

$$K_\nabla = \nabla \circ \nabla : S^{(1)} \to \Omega^2 \otimes S^{(1)}$$

of ∇ verifies $\overline{\pi} \circ K_\nabla = D \circ \nabla$.

If α stands for the projection $\Omega^2 \otimes S \to \Omega^2 \otimes J^{q-1}(E)$, one has

$$\alpha \circ (\overline{\pi} \circ K_\nabla) = D \circ D : S^{(1)} \to \Omega^2 \otimes J^{q-1}(E)$$

and consequently $\alpha \circ (\overline{\pi} \circ K_\nabla) = 0$. It follows that $\overline{\pi} \circ K_\nabla$ admits a well-defined and unique lift as a morphism of $\mathbb{C}$-sheaves:

$$K_S : S^{(1)} \to \Omega^2 \otimes \mathfrak{S}_S.$$

Since the first line of (6.18) is a complex, it is clear that the surjectivity of $S^{(2)} \to S^{(1)}$ implies the vanishing of the curvature K_S. From Theorem 6.3.1 applied to $S^{(1)}$, one deduces that $K_S \equiv 0$ if $S \simeq S^{(1)}$ is formally integrable.

Conversely, one can show that $K_S \equiv 0$ implies that S is regular and integrable. This non-trivial fact will not be proved here. Readers who are interested might consult [125].

Corollary 6.3.2. *If* $\mathfrak{S}_S^{(1)} = 0$ *and the natural morphism* $S^{(1)} \to S$ *is surjective, then* S *is integrable if and only if* $K_S \equiv 0$.

6.3.2   The Differential System $S_{\mathcal{W}}$

The theory sketched above will now be applied to a differential system which has integral forms of the abelian relations of a given planar web as solutions.

Let $\mathcal{W} = \mathcal{W}(\omega_1, \ldots, \omega_k)$ be a smooth k-web on $(\mathbb{C}^2, 0)$. It is harmless to assume that

$$\omega_i = dx + \theta_i \, dy$$

for $i \in \underline{k}$ and suitable germs $\theta_1, \ldots, \theta_k \in \mathcal{O}(\mathbb{C}^2, 0)$.

Let $X_i = \partial_y - \theta_i \, \partial_x$ and consider the following differential system

$$S_{\mathcal{W}} \quad : \qquad \begin{cases} \sum_{i=1}^k \varphi_i = 0, \\ X_i(\varphi_i) = 0 \quad i \in \underline{k} \, . \end{cases}$$

The space of holomorphic solutions of $S_{\mathcal{W}}$ will be denoted by $\mathrm{Sol}(\mathcal{W})$. Notice that $\mathrm{Sol}(\mathcal{W})$ fits into the exact sequence of vector spaces

$$0 \to \mathcal{A}(\mathcal{W}) \xrightarrow{f_0} \mathrm{Sol}(\mathcal{W}) \longrightarrow \mathbb{C}^k \xrightarrow{\Sigma} \mathbb{C} \to 0,$$

where the arrow from $\mathrm{Sol}(\mathcal{W})$ to $\mathbb{C}^k$ is given by evaluation at the origin. In particular the rank of $\mathcal{W}$ is maximal if and only if $\mathrm{Sol}(\mathcal{W})$ has dimension $k(k-1)/2$.

Notice that $S_{\mathcal{W}}$ is a differential linear system of first order and as such can be defined as follows. If E (resp. F) is the trivial bundle of rank k (resp. $k + 3$) over $(\mathbb{C}^2, 0)$, then $\mathrm{Sol}(\mathcal{W})$ can be identified with the sections of E with first jet belonging to the kernel of the map

$$\Phi = (F_{00}, F_{10}, F_{01}, G_1, \ldots, G_k) : J^1(E) \longrightarrow F$$

where

$$F_\sigma = \sum_{i=1}^k p_\sigma^i \qquad \text{and} \qquad G_i = p_{01}^i - \theta_i \, p_{10}^i$$

for σ such that $|\sigma| \leq 1$ and $i \in \underline{k}$.

The smoothness of $\mathcal{W}$ promptly implies that Φ has constant rank. The subbundle $\mathrm{Ker}(\Phi) \subset J^1(E)$ will be identified with $S_{\mathcal{W}}$.

Prolongations of $S_\mathcal{W}$

For every $\ell \geq 0$, denote by $S_\mathcal{W}^{(\ell)} \subset J^{1+\ell}(E)$ the ℓ-th prolongation of $S_\mathcal{W}$. To write down $S_\mathcal{W}^{(\ell)}$ explicitly, set

$$D^\tau = D_x^{\tau_1} \circ D_y^{\tau_2}$$

for $\tau = (\tau_1, \tau_2) \in \mathbb{N}^2$, where D_x and D_y stand for the total derivatives

$$D_x = \frac{\partial}{\partial x} + \sum_{i,\sigma} p_{\sigma(1)}^i \frac{\partial}{\partial p_\sigma^i} \qquad \text{and} \qquad D_y = \frac{\partial}{\partial y} + \sum_{i,\sigma} p_{\sigma(2)}^i \frac{\partial}{\partial p_\sigma^i}$$

(recall that $\sigma(1) = (\sigma_1 + 1, \sigma_2)$ and $\sigma(2) = (\sigma_1, \sigma_2 + 1)$ for every $\sigma = (\sigma_1, \sigma_2) \in \mathbb{N}^2$).

Using the notation just introduced, it is not hard to check that $S_\mathcal{W}^{(\ell)}$ is defined by the linear differential equations

$$D^\tau(F_\sigma) = 0, \quad |\sigma| = 0, 1 \qquad \text{and} \qquad D^\tau(G_i) = 0, \quad i \in \underline{k}$$

with $\tau \in \mathbb{N}^2$ satisfying $|\tau| \leq \ell$.

Notice that for every $\tau, \sigma \in \mathbb{N}^2$ and every $i \in \underline{k}$, the identities

$$D^\tau(F_\sigma) = \sum_{i=1}^{d} p_{\sigma+\tau}^i \qquad \text{and} \qquad D^\tau(G_i) = p_{\tau(2)}^i - D^\tau\left(\theta_i \, p_{10}^i\right)$$

hold true. Since D_x and D_y are derivations it follows that

$$D^\tau(G_i) = p_{\tau(2)}^i - \sum_{\alpha+\beta=\tau} D^\alpha(\theta_i) D^\beta(p_{10}^i)$$

$$= p_{\tau(2)}^i - \sum_{\alpha+\beta=\tau} D^\alpha(\theta_i) \, p_{\beta(1)}^i$$

$$= p_{\tau(2)}^i - \theta_i \, p_{\tau(1)}^i + \sum_{\kappa=1}^{|\tau|-1} L_\kappa^\tau(i)$$

where

$$L_\kappa^\tau(i) = \sum_{\substack{|\beta|=\kappa \\ \alpha+\beta=\tau}} D^\alpha(\theta_i) \, p_{\beta(1)}^i = \sum_{\substack{|\beta|=\kappa \\ \alpha+\beta=\tau}} \left(\frac{\partial^{|\alpha|} \theta_i}{\partial x^{\alpha_1} \partial y^{\alpha_2}}\right) p_{\beta(1)}^i \, .$$

If $\ell \geq 0$ and if $(e_1, \ldots, e_k)$ stands for a basis of E, then $\mathfrak{S}^{(\ell)}$, the ℓ-th symbol of $S_{\mathcal{W}}$, is generated by the elements

$$\sum_{\substack{i=1,\ldots,k \\ |\sigma|=1+\ell}} \xi_\sigma^i \left(dx^{\sigma_1} \cdot dy^{\sigma_2}\right) \otimes e_i \in \mathrm{Sym}^{1+\ell}\left(T^*(\mathbb{C}^2, 0)\right) \otimes E$$

subject to the conditions

$$(1)_{ab} \qquad \sum_{i=1}^k \xi_{ab}^i = 0 \qquad \text{and} \qquad (2)_{ab}^i \qquad \xi_{a,b+1}^i = \theta_i \, \xi_{a+1,b}^i$$

for every $i \in \underline{k}$ and every $(a, b) \in \mathbb{N}^2$ satisfying $a + b = 1 + \ell$.
The equations $(2)_{ab}^i$ are equivalent to the following ones:

$$(2')_{ab}^i \qquad\qquad \xi_{ab}^i = (\theta_i)^b \, \xi_{\ell+1,0}^i.$$

Consequently $\mathfrak{S}^{(\ell)}$ is generated by elements of the form

$$\sum_{i=1}^k z^i \left(\sum_{a+b=1+\ell} (\theta_i)^b \left(dx^a \cdot dy^b\right)\right) \otimes e_i$$

where the z_i are subject to the conditions

$$(1')_s \qquad\qquad \sum_{i=1}^k z^i \, (\theta_i)^s = 0$$

for $s \in \{0, \ldots, \ell + 1\}$. These equations can be written in matrix form:

$$\left(V_\theta^{(\ell)}\right) \qquad\qquad \begin{pmatrix} 1 & \cdots & 1 \\ \theta_1 & \cdots & \theta_k \\ \vdots & & \vdots \\ (\theta_1)^{\ell+1} & \cdots & (\theta_k)^{\ell+1} \end{pmatrix} \begin{pmatrix} z^1 \\ \vdots \\ z^k \end{pmatrix} = 0.$$

Because $\mathcal{W}$ is smooth, the functions θ_i have pairwise distinct values at every point of $(\mathbb{C}^2, 0)$. Thus $\left(V_\theta^{(\ell)}\right)$ is a linear system of Vandermonde type and there are only two possibilities.

- If $\ell \geq k - 2$, then $\left(V_\theta^{(\ell)}\right)$ has no non-trivial solution; in other terms, $\mathfrak{S}^{(\ell)} = (0)$.
- If $\ell \in \{0, \ldots, k - 3\}$, then $\left(V_\theta^{(\ell)}\right)$ is a linear system of rank $\ell + 2$, therefore $\mathfrak{S}^{(\ell)}$ has dimension $k - 2 - \ell$ and can be parameterized explicitly via the inverse of

the Vandermonde matrix associated with the functions θ_i. If $(\alpha_{ij})_{i,j=1}^k$ stands for the inverse of the Vandermonde matrix $((\theta_i)^{j-1})_{i,j=1}^k$ then

$$\mathfrak{S}^{(\ell)} = \left\langle \sum_{i=1}^k \alpha_{im} \left(\sum_{a+b=1+\ell} (\theta_i)^b \, dx^a \cdot dy^b \right) \otimes e_i \,\middle|\, m = \ell + 3, \ldots, k \right\rangle.$$

6.3.3 Pantazi–Hénaut Criterion

Set $\mathscr{S}_\mathcal{W}$ as the differential system $S_\mathcal{W}^{(k-3)}$. According to Sect. 6.3.2 it has the following properties:

1. It is a subbundle of $J^{k-2}(E)$ of rank $k(k-1)/2$;
2. Its symbol, denoted by $\sigma_\mathcal{W}$, is a sub-line bundle of $\mathrm{Sym}^{k-2}(T^*(\mathbb{C}^2, 0)) \otimes E$;
3. The map $\mathscr{S}_\mathcal{W}^{(1)} \to \mathscr{S}_\mathcal{W}$ is an isomorphism (thus $\sigma_\mathcal{W}^{(1)} = 0$).

The discussion laid down in Sect. 6.3.1 implies the existence of **Hénaut's connection** of $\mathcal{W}$

$$\nabla_\mathcal{W} : \mathscr{S}_\mathcal{W}^{(1)} \to \Omega^1 \otimes \mathscr{S}_\mathcal{W}^{(1)}.$$

Its curvature is a $\mathcal{O}$-linear operator

$$\Theta_\mathcal{W} : \mathscr{S}_\mathcal{W}^{(1)} \to \Omega^2 \otimes \sigma_\mathcal{W}.$$

called the **Pantazi–Hénaut curvature** of the web $\mathcal{W}$.

Theorem 6.3.3. *The following assertions are equivalent.*

1. *$\mathcal{W}$ has maximal rank;*
2. *Pantazi–Hénaut curvature $\Theta_\mathcal{W}$ vanishes identically.*

The above theorem can be understood as a generalization from 3-webs to arbitrary k-webs of the equivalence between the items (**b**) and (**c**) of Theorem 1.2.4.

If the functions $\theta_1, \ldots, \theta_k$ are given explicitly, one can easily construct an effective algorithm (see [107, Appendice] for an implementation in MAPLE) which computes the curvature $\Theta_\mathcal{W}$ as defined above. If instead of the functions θ_i one only knows an explicit expression of a k-symmetric 1-form defining $\mathcal{W}$, then the implicit approach implemented by Hénaut shows the existence of an algorithm to compute $\Theta_\mathcal{W}$. Such algorithm is not as easy to derive as in the approach followed here. Albeit, Ripoll spelled out and implemented in MAPLE the corresponding algorithm for 3, 4, and 5-webs presented in implicit form.

An extensive study of Hénaut's connection $\nabla_\mathcal{W}$ remains to be done. The authors believe that a careful investigation of its properties may lead to an answer to Problem 1. Casale's results [25] on the Galois–Malgrange groupoid of codimension one foliations could be useful in this context.

6.3.4 Mihăileanu's Criterion

Despite the relative ease of implementing the computation of $\Theta_{\mathcal{W}}$, it seems difficult to interpret the corresponding matrix of 2-forms. Nevertheless Mihăileanu, building on Pantazi's result and after ingenious computations, shows in [87] that

$$K(\mathcal{W}) = \sum_{\mathcal{W}_3 < \mathcal{W}} K(\mathcal{W}_3),$$

the sum of curvatures of all 3-subwebs of $\mathcal{W}$, appears as a linear combination of the coefficients of Pantazi–Hénaut curvature. From this result he obtained from Theorem 6.3.3 the following necessary condition for rank maximality.

Theorem 6.3.4 (Mihăileanu's Criterion). *If $\mathcal{W}$ is a planar k-web of maximal rank then $K(\mathcal{W})$ vanishes identically.*

By definition, the 2-form $K(\mathcal{W})$ is the **curvature of** $\mathcal{W}$. An intrinsic interpretation of it has been recently provided in [118, Théorème 5.2] when $k \leq 6$, and stated for arbitrary k in [78, p. 281], [120]. According to these references, the curvature $K(\mathcal{W})$ is nothing else than the trace of Pantazi–Hénaut curvature $\Theta_{\mathcal{W}}$.

As far as we know, there is no written proof of Mihăileanu's criterion. This is quite regrettable since it is a key ingredient in several recent works on planar webs (such as [90, 106]).

For linear webs, Mihăileanu's criterion takes a very nice form. Indeed, by a direct computation, one can verify that if $\mathcal{W} = \mathcal{W}(dx + e_1 dy, \ldots, dx + e_k dy)$ is a linear web, one has $K_{\mathcal{W}} = \sum_{i=1}^{k} \partial_{xx}^2(e_i)$. Since $\sum_{i=1}^{k} \partial_{xx}^2(e_i) \equiv 0$ is exactly the condition needed for a certain kind of converse to Abel's theorem applies (see [135] or [74]), one deduces the following result:

Theorem 6.3.5. *Assume that $\mathcal{W}$ is linearizable. The following assertions are equivalent:*

- *$\mathcal{W}$ is algebraizable;*
- *$\mathcal{W}$ is of maximal rank;*
- *$\mathcal{W}$ carries a complete abelian relation;*
- *$K_{\mathcal{W}}$ vanishes identically.*

Since being linearizable can be characterized in terms of differential invariants (cf. Sect. 6.1.3 above), the preceding result gives an invariant characterization of algebraizable planar webs.

6.4 Classification of CDQL-Webs

Although Pantazi–Hénaut criterion together with the linearization criterion presented in Sect. 6.1 provide an effective algorithmic way to decide if a given planar web is exceptional or not, the classification problem for these objects is wide open to-day. The only classification results available so far concern only the classification of (very) particular classes of webs. Still worse, only two classes have been studied so far. The first class consists of the germs of 5-webs on $(\mathbb{C}^2, 0)$ of the form $\mathcal{W}(x, y, x + y, x - y, u(x) + v(y))$ mentioned in Example 2.2.4 of Chap. 2. The second class of exceptional webs so far classified consists of the Completely Decomposable Quasi-Linear webs ('CDQL-webs' for short) on compact complex surfaces.

6.4.1 Definition

Linear webs are classically defined as the ones for which all the leaves are open subsets of lines. Here we will adopt the following global definition. A web $\mathcal{W}$ on a compact complex surface S is **linear** if

- the universal covering of S is an open subset $\tilde{S}$ of $\mathbb{P}^2$;
- the group of deck transformations acts on $\tilde{S}$ by automorphisms of $\mathbb{P}^2$; and
- the pull-back of $\mathcal{W}$ to $\tilde{S}$ is linear in the classical sense.

By definition, a **CDQL** $(k + 1)$**-web** on a compact complex surface S is the superposition of k linear foliations and one non-linear foliation. It can be verified (see [106]) that the only compact complex surfaces carrying CDQL $(k + 1)$-webs when $k \geq 2$ are the projective plane, the complex tori, and the Hopf surfaces. Moreover the only Hopf surfaces admitting four distinct linear foliations are the primary Hopf surfaces H_α, $|\alpha| < 1$. Here H_α is the quotient of $\mathbb{C}^2 \setminus \{0\}$ by the map $(x, y) \mapsto (\alpha x, \alpha y)$.

The linear foliations on complex tori are pencils of parallel lines on their universal coverings. The ones on Hopf surfaces H_α are either pencils of parallel lines or the pencil of lines through the origin of $\mathbb{C}^2$. In particular a completely decomposable linear web on a compact complex surface is algebraic[4] on its universal covering.

6.4.2 On the Projective Plane

The classification of exceptional CDQL webs on the projective plane is summarized in the following result.

Theorem 6.4.1. *Up to projective automorphisms, there are exactly four infinite families and thirteen sporadic exceptional CDQL webs on* $\mathbb{P}^2$.

---

[4]Beware that algebraic here means that they are locally dual to plane curves. In the cases under scrutiny they are dual to products of lines.

To describe the exceptional webs mentioned in Theorem 6.4.1, the notation of [106] will be adopted: if ω is a rational k-symmetric differential 1-form, then $[\omega]$ denotes the k-web on $\mathbb{P}^2$ induced by it.

In suitable affine coordinates $(x, y) \in \mathbb{C}^2 \subset \mathbb{P}^2$, the four infinite families are

$$\mathcal{A}_I^k = \left[(dx^k - dy^k)\right] \boxtimes \left[d(xy)\right] \qquad \text{where } k \geq 4;$$

$$\mathcal{A}_{II}^k = \left[(dx^k - dy^k)(xdy - ydx)\right] \boxtimes \left[d(xy)\right] \qquad \text{where } k \geq 3;$$

$$\mathcal{A}_{III}^k = \left[(dx^k - dy^k)\, dx\, dy\right] \boxtimes \left[d(xy)\right] \qquad \text{where } k \geq 2;$$

$$\mathcal{A}_{IV}^k = \left[(dx^k - dy^k)\, dx\, dy\, (xdy - ydx)\right] \boxtimes \left[d(xy)\right] \qquad \text{where } k \geq 1.$$

The diagram below shows how these webs relate to each other in terms of inclusions for a fixed k. Moreover if k divides k' then $\mathcal{A}_I^k, \mathcal{A}_{II}^k, \mathcal{A}_{III}^k, \mathcal{A}_{IV}^k$ are subwebs of $\mathcal{A}_I^{k'}, \mathcal{A}_{II}^{k'}, \mathcal{A}_{III}^{k'}, \mathcal{A}_{IV}^{k'}$, respectively.

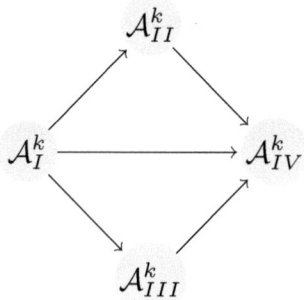

All the webs above are invariant by the action $t \cdot (x, y) = (tx, ty)$ of $\mathbb{C}^*$ on $\mathbb{P}^2$. Among the thirteen sporadic examples of exceptional CDQL webs on the projective plane, seven (four 5-webs, two 6-webs, and one 7-web) are also invariant by the same $\mathbb{C}^*$-action. They are:

$$\mathcal{A}_5^a = \left[dx\, dy\, (dx + dy)(xdy - ydx)\right] \boxtimes \left[d(xy(x + y))\right];$$

$$\mathcal{A}_5^b = \left[dx\, dy\, (dx + dy)(xdy - ydx)\right] \boxtimes \left[d\left(\frac{xy}{x+y}\right)\right];$$

$$\mathcal{A}_5^c = \left[dx\, dy\, (dx + dy)(xdy - ydx)\right] \boxtimes \left[d\left(\frac{x^2 + xy + y^2}{xy\,(x+y)}\right)\right];$$

$$\mathcal{A}_5^d = \left[dx\, (dx^3 + dy^3)\right] \boxtimes \left[d\left(x(x^3 + y^3)\right)\right];$$

$$\mathcal{A}_6^a = \left[dx\, (dx^3 + dy^3)(xdy - ydx)\right] \boxtimes \left[d\left(x(x^3 + y^3)\right)\right];$$

$$\mathcal{A}_6^b = \left[dx\, dy\, (dx^3 + dy^3)\right] \boxtimes \left[d(x^3 + y^3)\right];$$

$$\mathcal{A}_7 = \left[dx\, dy\, (dx^3 + dy^3)(xdy - ydx)\right] \boxtimes \left[d(x^3 + y^3)\right].$$

Four of the remaining six sporadic exceptional CDQL webs (one k-web for each $k \in \{5, 6, 7, 8\}$) share the same non-linear foliation $\mathcal{F}$: the pencil of conics through four points in general position. They all have been previously known (see [119]).

$$\mathcal{B}_5 = \left[dx\, dy\, d\left(\tfrac{x}{1-y}\right) d\left(\tfrac{y}{1-x}\right) \right] \boxtimes \left[d\left(\tfrac{xy}{(1-x)(1-y)}\right) \right];$$
$$\mathcal{B}_6 = \mathcal{B}_5 \qquad\qquad\qquad\quad \boxtimes \left[d\left(x+y\right) \right];$$
$$\mathcal{B}_7 = \mathcal{B}_6 \qquad\qquad\qquad\quad \boxtimes \left[d\left(\tfrac{x}{y}\right) \right];$$
$$\mathcal{B}_8 = \mathcal{B}_7 \qquad\qquad\qquad\quad \boxtimes \left[d\left(\tfrac{1-x}{1-y}\right) \right].$$

The last two sporadic CDQL exceptional webs (the 5-web $\mathcal{H}_5$ and the 10-web $\mathcal{H}_{10}$) of Theorem 6.2.2 also share the same non-linear foliation: the **Hesse pencil of cubics**. Recall that this pencil is the one generated by a smooth cubic and its Hessian and that it is unique up to automorphisms of $\mathbb{P}^2$. Explicitly (with $\xi_3 = \exp(2i\pi/3)$):

$$\mathcal{H}_5 = \left[(dx^3 + dy^3)\, d\left(\tfrac{x}{y}\right) \right] \boxtimes \left[d\left(\tfrac{x^3+y^3+1}{xy}\right) \right]$$

$$\mathcal{H}_{10} = \left[(dx^3 + dy^3) \prod_{k=0}^{2} \left(d\left(\tfrac{y-\xi_3^k}{x}\right) \cdot d\left(\tfrac{x-\xi_3^k}{y}\right) \right) \right] \boxtimes \left[d\left(\tfrac{x^3+y^3+1}{xy}\right) \right].$$

The 10-web $\mathcal{H}_{10}$ is better described synthetically: it is the superposition of the Hesse pencil of cubics and of the nine pencils of lines with base points at the base points of Hesse pencil. It shares a number of features with Bol's web $\mathcal{B}_5$. For instance, they both have a huge group of birational automorphisms (the symmetric group $\mathfrak{S}_5$ for $\mathcal{B}_5$ and Hesse's group G_{216} for $\mathcal{H}_{10}$), and their abelian relations can be expressed in terms of logarithms and dilogarithms.

Because they have parallel 4-subwebs whose slopes have non-real cross-ratio the webs $\mathcal{A}_{III}^k$ for $k \geq 3$, $\mathcal{A}_{IV}^k$ for $k \geq 3$, $\mathcal{A}_5^d, \mathcal{A}_6^a, \mathcal{A}_6^b$, and $\mathcal{A}_7$ do not admit real models. The web $\mathcal{H}_{10}$ does not admit a real model either. There are a number of ways to verify this fact. One possibility is to observe that the lines passing through two of the nine base points always contain a third and notice that this contradicts Sylvester–Gallai Theorem [38]: for every finite set of non-collinear points in $\mathbb{P}^2_{\mathbb{R}}$ there exists a line containing exactly two points of the set. All the other exceptional CDQL webs on the projective plane admit real models. For a sample see Fig. 6.1.

6.4.3   On Hopf Surfaces

The classification of exceptional CDQL webs on Hopf surfaces reduces to the one on $\mathbb{P}^2$. The result is not particularly interesting. One has just to remark that a foliation on a Hopf surface of type H_α when lifted to $\mathbb{C}^2 \setminus \{0\}$ gives rise to an algebraic foliation on $\mathbb{C}^2$ invariant by the $\mathbb{C}^*$-action $t \cdot (x, y) = (tx, ty)$, and then use the classification of exceptional CDQL webs on $\mathbb{P}^2$ to obtain the following result.

Fig. 6.1 A sample of the real models for exceptional CDQL on $\mathbb{P}^2$. In the *first and second rows* the first three members of the infinite family $\mathcal{A}_I^k$ and $\mathcal{A}_{II}^k$, respectively. In the *third row*, from *left to right*, $\mathcal{A}_{III}^2$, $\mathcal{A}_{IV}^1$ and $\mathcal{A}_{IV}^2$. In the *fourth row*: $\mathcal{A}_5^a$, $\mathcal{A}_5^b$ and $\mathcal{A}_5^c$

Corollary 6.4.2. *Up to automorphisms, the only exceptional CDQL webs on Hopf surfaces are the quotients of the restrictions of the webs $\mathcal{A}_*^*$ to $\mathbb{C}^2 \setminus \{0\}$ by the group of deck transformations.*

6.4.4   On Abelian Surfaces

The CDQL webs on tori are the superpositions of a non-linear foliation with a product of foliations induced by global holomorphic 1-forms. Since étale coverings between complex tori abound and because the pull-backs of exceptional CDQL webs under these are still exceptional CDQL webs, it is natural to extend the notion of isogenies between complex tori to isogenies between webs on tori.

Two webs $\mathcal{W}_1, \mathcal{W}_2$ on complex tori T_1, T_2 are **isogeneous** if there exist a complex torus T and two étale morphisms $\pi_i : T \to T_i$ for $i = 1, 2$, such that $\pi_1^*(\mathcal{W}_1) = \pi_2^*(\mathcal{W}_2)$.

Theorem 6.4.3. *Up to isogenies, there are exactly three sporadic exceptional CDQL k-webs (one for each $k \in \{5, 6, 7\}$) and one continuous family of exceptional CDQL 5-webs on complex tori.*

The elements of the continuous family are

$$\mathcal{E}_\tau = \left[dx \, dy \, (dx^2 - dy^2) \right] \boxtimes \left[d \left(\frac{\vartheta_1(x, \tau)\vartheta_1(y, \tau)}{\vartheta_4(x, \tau)\vartheta_4(y, \tau)} \right)^2 \right].$$

respectively defined on the square $(E_\tau)^2$ of the elliptic curve $E_\tau = \mathbb{C}/(\mathbb{Z} \oplus \mathbb{Z}\tau)$ for $\tau \in \mathbb{H} = \{z \in \mathbb{C} \mid \Im m(z) > 0\}$ arbitrary. The functions ϑ_i involved in the definition are the classical **Jacobi theta functions**, whose definition is now recalled.

For $(\mu, \nu) \in \{0, 1\}^2$ and $\tau \in \mathbb{H}$ fixed, let $\vartheta_{\mu,\nu}(\cdot, \tau)$ be the entire function on $\mathbb{C}$

$$\vartheta_{\mu,\nu}(x, \tau) = \sum_{n=-\infty}^{+\infty} (-1)^{n\nu} \exp\left(i\,\pi\left(n + \frac{\mu}{2}\right)^2 \tau + 2\,i\,\pi\left(n + \frac{\mu}{2}\right)x \right).$$

These are usually called the **theta functions with characteristic**. The Jacobi theta functions ϑ_i are nothing else than

$$\vartheta_1 = -i\,\vartheta_{1,1}, \quad \vartheta_2 = \vartheta_{1,0}, \quad \vartheta_3 = \vartheta_{0,0} \quad \text{and} \quad \vartheta_4 = \vartheta_{0,1}.$$

The webs $\mathcal{E}_\tau$ first appeared in Buzano's work [24] but their rank was not determined at that time. They were later rediscovered[5] in [113] where it is proved that they all are exceptional and that $\mathcal{E}_\tau$ is isogeneous to $\mathcal{E}_{\tau'}$ if and only if τ and τ' belong to the same orbit under the natural action of the $\mathbb{Z}/2\mathbb{Z}$ extension of $\Gamma_0(2) \subset PSL(2, \mathbb{Z})$ generated by $\tau \mapsto -2\tau^{-1}$. Thus the continuous family of exceptional CDQL webs on tori is parameterized by a $\mathbb{Z}/2\mathbb{Z}$-quotient of the modular curve $X_0(2)$.

---

[5]These are the 5-webs mentioned in Example 2.2.4.

The sporadic CDQL 7-web $\mathcal{E}_7$ is related to a particular element of the previous family. Indeed $\mathcal{E}_7$ is the 7-web on $(E_{1+i})^2$

$$\mathcal{E}_7 = \left[dx^2 + dy^2\right] \boxtimes \mathcal{E}_{1+i} .$$

The sporadic CDQL 5-web $\mathcal{E}_5$ lives naturally on $(E_{\xi_3})^2$ and can be described as the superposition of the linear web

$$\left[dx \, dy \, (dx - dy) \, (dx + (\xi_3)^2 \, dy)\right]$$

and of the non-linear foliation

$$\left[d\left(\frac{\vartheta_1(x, \xi_3)\vartheta_1(y, \xi_3)\vartheta_1(x - y, \xi_3)\vartheta_1\left(x + (\xi_3)^2 \, y, \xi_3\right)}{\vartheta_2(x, \xi_3)\vartheta_3(y, \xi_3)\vartheta_4(x - y, \xi_3)\vartheta_3\left(x + (\xi_3)^2 \, y, \xi_3\right)}\right)\right] .$$

The sporadic CDQL 6-web $\mathcal{E}_6$ also lives on $(E_{\xi_3})^2$ and is best described in terms of Weierstrass $\wp$-function.

$$\mathcal{E}_6 = \left[\, dx \, dy \, (dx^3 + dy^3)\right] \boxtimes \left[\wp(x, \xi_3)^{-1}dx + \wp(y, \xi_3)^{-1}dy\right].$$

For a more geometric description of these exceptional *elliptic webs* the reader is invited to consult [106, Section 4].

6.4.5   Outline of the Proof

In the remaining of this section, the proof of Theorem 6.4.1 will be sketched. It makes use of Mihăileanu's criterion in an essential way. Its starting point is the following trivial observation: if $K(\mathcal{W})$, the curvature of $\mathcal{W}$, is zero then it must be, in particular, a holomorphic 2-form.

Regularity of the Curvature

The tautology just spelled out makes clear the relevance of obtaining criterion to ensure the absence of poles of $K(\mathcal{W})$. The result obtained in this respect in [106] is best stated in terms of $\beta_{\mathcal{F}}(\mathcal{W})$—the $\mathcal{F}$-**barycenter** of a web $\mathcal{W}$. If $\mathcal{W}$ is a k-web and if $\mathcal{F}$ is a foliation not contained in $\mathcal{W}$, both defined on a surface S, then at a generic point $p \in S$ the tangents of $\mathcal{F}$ and $\mathcal{W}$ determine $k + 1$ points in $\mathbb{P}(T_pS)$. The complement of the point $[T_p\mathcal{F}]$ in $\mathbb{P}(T_pS)$ is clearly isomorphic to $\mathbb{C}$ and any two distinct isomorphisms differ by an affine map. The $\mathcal{F}$-barycenter of $\mathcal{W}$ is then defined as the foliation on S with tangent at a generic point p of S determined

Fig. 6.2 The $\mathcal{L}$-barycenter of
a CDL web $\mathcal{W}$

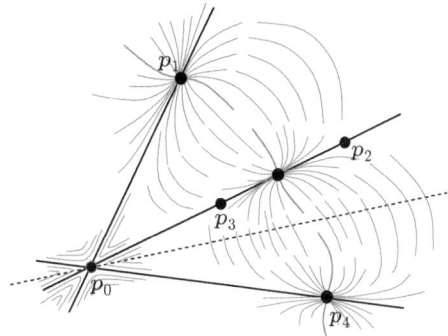

by the barycenter of the k tangents directions of $\mathcal{W}$ at p in the affine structure on
$\mathbb{C} \simeq \mathbb{P}(T_p S) \setminus [T_p \mathcal{F}]$ determined by $\mathcal{F}$.

Theorem 6.4.4. *Let $\mathcal{F}$ be a foliation and $\mathcal{W} = \mathcal{F}_1 \boxtimes \cdots \boxtimes \mathcal{F}_k$ be a k-web, $k \geq$
2, both defined on the same domain $U \subset \mathbb{C}^2$. Suppose that C is an irreducible
component of $\mathrm{tang}(\mathcal{F}, \mathcal{F}_1)$ that is not contained in $\Delta(\mathcal{W})$. The curvature $K(\mathcal{F} \boxtimes \mathcal{W})$
is holomorphic over a generic point of C if and only if the curve C is $\mathcal{F}$-invariant
or $\beta_{\mathcal{F}}(\mathcal{W}')$-invariant, where $\mathcal{W}' = \mathcal{F}_2 \boxtimes \cdots \boxtimes \mathcal{F}_k$.*

This result is the key tool used in [106] to achieve the classification. Having it at
hand, the next step is to the describe the $\mathcal{L}$-barycenters of Completely Decomposable Linear webs (**CDL-webs** for short) with respect to a linear foliation $\mathcal{L}$.

$\mathcal{L}$-Barycenters of CDL Webs

A linear foliation $\mathcal{L}$ on $\mathbb{P}^2$ is nothing else than a pencil of lines. Thus, it is determined
by its unique singular point. So, let $p_0 \in \mathbb{P}^2$ be the point determining $\mathcal{L}$ and let
$\{p_1, \ldots, p_k\} \subset \mathbb{P}^2$ be the set of points determining a CDL k-web $\mathcal{W}$. In order to
describe the $\mathcal{L}$-barycenter of $\mathcal{W}$, let $\Pi : S \to \mathbb{P}^2$ be the blow-up of $\mathbb{P}^2$ at p_0; E its
exceptional divisor; $\pi : S \to \mathbb{P}^1$ be the fibration on S induced by the lines through
p_0; $\mathcal{G}$ be the foliation $\Pi^* \beta_{\mathcal{L}}(\mathcal{W})$; and ℓ_i be the strict transform of the line $\overline{p_0 p_i}$
under Π for $i = 1, \ldots, k$.

If the points $\{p_0, \ldots, p_k\}$ are aligned, then $\beta_{\mathcal{F}}(\mathcal{W})$ is also a pencil of lines whose
base point is the p_0-barycenter of $\{p_1, \ldots, p_k\}$. If instead the points $\{p_0, \ldots, p_k\}$
are not aligned, then a simple computation shows that the foliation $\mathcal{G}$ is a Riccati
foliation with respect to π, that is, $\mathcal{G}$ has no tangency with the generic fiber of π
(Fig. 6.2).

In fact, a much more precise description of $\mathcal{G}$ can be obtained. In [106,
Lemma 6.1] it is shown that this foliation has the following properties:

1. the exceptional divisor E of Π is $\mathcal{G}$-invariant;
2. the only $\mathcal{G}$-invariant fibers of π are the lines ℓ_i, for $i \in \underline{k}$;

3. the singular set of $\mathcal{G}$ is contained in the union $\bigcup_{i \in \underline{k}} \ell_i$;
4. over each line ℓ_i, the foliation $\mathcal{G}$ has two singularities. One is a complex saddle at the intersection of ℓ_i with E, the other is a complex node at the p_0-barycenter of $\mathcal{P}_i = \{p_1, \ldots, p_k\} \cap \ell_i$. Moreover, if r_i is the cardinality of $\mathcal{P}_i$, then the quotient of the eigenvalues of the saddle (resp. node) over ℓ_i is $-r_i/k$ (resp. r_i/k);
5. the monodromy of $\mathcal{G}$ around ℓ_i is finite and of order $k/\gcd(k, r_i)$;
6. the separatrices of $\beta_{\mathcal{F}}(\mathcal{W})$ through p_0 are the lines $\overline{p_0 p_i}$, $i \in \underline{k}$.

It is interesting to notice that the generic leaf of $\beta_{\mathcal{F}}(\mathcal{W})$ is transcendental in general. Indeed, the cases when there are more algebraic leaves than the obvious ones (the lines $\overline{p_0 p_i}$) are conveniently characterized by [106, Proposition 6.1], which says that the foliation $\beta_{\mathcal{F}}(\mathcal{W})$ has an algebraic leaf distinct from the lines $\overline{p_0 p_i}$ if and only if all its singularities distinct from p_0 are aligned. Moreover if this is the case then all its leaves are algebraic.

The ℓ-Polar Map and Bounds for the Degree of $\mathcal{F}$

For a set $\mathcal{P}$ of k distinct points in $\mathbb{P}^2$, let $\mathcal{W}(\mathcal{P})$ be the CDL k-web on the projective plane formed by the pencils of lines with base points at the points of $\mathcal{P}$.

Once the description of the $\mathcal{L}$-barycenters of CDL webs has been laid down, the next step is to use it to obtain constraints on the non-linear foliation $\mathcal{F}$ and on the position of the points $\mathcal{P}$ in case $\mathcal{W}(\mathcal{P}) \boxtimes \mathcal{F}$ has zero curvature.

It is not hard to show (see [106, Section 8]) that when the cardinality of $\mathcal{P}$ is at least 4, either (a) there are three aligned points in $\mathcal{P}$; or (b) $\mathcal{P}$ is a set of 4 points in general position and $\mathcal{F}$ is the pencil of conics through them.

When in case (b) there is not much left to do, since $\mathcal{F} \boxtimes \mathcal{W}(\mathcal{P})$ is nothing more than Bol's 5-web; in case (a) one is naturally led to consider a line ℓ containing k_ℓ points of $\mathcal{P}$, with $k_\ell \geq 3$; and the pencil $V = \{\text{tang}(\mathcal{F}, \mathcal{L}_p)\}_{p \in \ell}$ of polar curves of $\mathcal{F}$ centered at ℓ. It can be shown that ℓ is a fixed component of V (in other words, ℓ is $\mathcal{F}$-invariant); and the restriction of $V - \ell$ to ℓ defines a non-constant rational map $f : \ell \simeq \mathbb{P}^1 \to \mathbb{P}^1$. The map f is characterized by the following equalities between divisors on ℓ

$$f^{-1}(p) = \left(\text{tang}(\mathcal{F}, \mathcal{L}_p) - \ell \right)\Big|_\ell,$$

with $p \in \ell$ arbitrary. The map f is called the ℓ-**polar map of** $\mathcal{F}$.

Once all these properties of f are settled, it follows from a simple application of Riemann–Hurwitz formula that the degree of $\mathcal{F}$ is at most four. Moreover, if $\deg(\mathcal{F}) \geq 2$, then $k_\ell \leq 7 - \deg(\mathcal{F})$.

The Final Steps

At this point the proof has a two-fold ramification. In one branch one is led to consider foliations of degree one and put to a good use the acquired knowledge on the structure of the space of abelian relations of web admitting infinitesimal automorphisms, see [106, Section 9]. In the other branch, one derives from the structure of the $\mathcal{L}$-barycenter of CDL webs, the normal forms for the ℓ-polar map of $\mathcal{F}$ presented in [106, Table 1]. Then, the proof goes by a case-by-case analysis, see [106, Section 10]. While the arguments can be considered elementary, they are too involved to be detailed here.

6.5   Further Examples

In this last section, different examples of exceptional webs are listed. Apart from the fact that they all are exceptional webs, there is no general directrix. The reason behind this chaotic exposition is the lack of a general framework encompassing all known exceptional webs.

6.5.1   Polylogarithmic Webs

It is well known that the **polylogarithms**

$$\mathrm{Li}_n(z) = \sum_{k=1}^{\infty} \frac{z^k}{k^n}$$

satisfy functional equations in two variables, at least when n is small, see [83].

For instance, Spence and Kummer have independently established some variants of the following functional equation, nowadays called **Spence–Kummer equation**, satisfied by the trilogarithm Li_3

$$2\,\mathrm{Li}_3(x) + 2\,\mathrm{Li}_3(y) - \mathrm{Li}_3\left(\frac{x}{y}\right) + 2\,\mathrm{Li}_3\left(\frac{1-x}{1-y}\right) + 2\,\mathrm{Li}_3\left(\frac{x(1-y)}{y(1-x)}\right)$$
$$-\mathrm{Li}_3(xy) + 2\,\mathrm{Li}_3\left(\frac{x(y-1)}{(1-x)}\right) + 2\,\mathrm{Li}_3\left(\frac{(y-1)}{y(1-x)}\right) - \mathrm{Li}_3\left(\frac{x(1-y)^2}{y(1-x)^2}\right)$$
$$= 2\,\mathrm{Li}_3(1) - \log(y)^2 \log\left(\frac{1-y}{1-x}\right) + \frac{\pi^2}{3}\log(y) + \frac{1}{3}\log(y)^3$$

when x and y are real numbers subject to the constraint $0 < x < y < 1$.

Kummer proved that the tetralogarithm and the pentalogarithm verify similar equations. If $\zeta = 1 - x$ and $\eta = 1 - y$ with $x, y \in \mathbb{R}$ and $0 < x < y < 1$, then the tetralogarithm $\mathbf{Li}_4$ satisfies the equation $\mathcal{K}(4)$, written down below:

$$\mathbf{Li}_4\left(-\frac{x^2 y \eta}{\zeta}\right) + \mathbf{Li}_4\left(-\frac{y^2 x \zeta}{\eta}\right) + \mathbf{Li}_4\left(\frac{x^2 y}{\eta^2 \zeta}\right) + \mathbf{Li}_4\left(\frac{y^2 x}{\zeta^2 \eta}\right)$$

$$-6\,\mathbf{Li}_4(xy) - 6\,\mathbf{Li}_4\left(\frac{xy}{\eta \zeta}\right) - 6\,\mathbf{Li}_4\left(-\frac{xy}{\eta}\right) - 6\,\mathbf{Li}_4\left(-\frac{xy}{\zeta}\right)$$

$$-3\,\mathbf{Li}_4(x\eta) - 3\,\mathbf{Li}_4(y\zeta) - 3\,\mathbf{Li}_4\left(\frac{x}{\eta}\right) - 3\,\mathbf{Li}_4\left(\frac{y}{\zeta}\right)$$

$$-3\,\mathbf{Li}_4\left(-\frac{x\eta}{\zeta}\right) - 3\,\mathbf{Li}_4\left(-\frac{y\zeta}{\eta}\right) - 3\,\mathbf{Li}_4\left(-\frac{x}{\eta\zeta}\right) - 3\,\mathbf{Li}_4\left(-\frac{y}{\eta\zeta}\right)$$

$$+6\,\mathbf{Li}_4(x) + 6\,\mathbf{Li}_4(y) + 6\,\mathbf{Li}_4\left(-\frac{x}{\zeta}\right) + 6\,\mathbf{Li}_4\left(-\frac{y}{\zeta}\right)$$

$$= -\frac{3}{2}\log^2(\zeta)\log^2(\eta).$$

The pentalogarithm $\mathbf{Li}_5$ satisfies an equation of the same type, which will be referred to as $\mathcal{K}(5)$. It involves more than thirty terms and will not be written down to save space.

All the known functional equations, in two variables, satisfied by the classical polylogarithms $\mathbf{Li}_n$ are of the form

$$\sum_{i=1}^{N} c_i\,\mathbf{Li}_n(U_i) \equiv \mathbf{Elem}_n \qquad (6.19)$$

where $c_1, \ldots, c_N$ are integers; $U_1, .., U_N$ are rational functions; $\mathbf{Elem}_n$ is of the form $P(\mathbf{Li}_{k_1} \circ V_1, \ldots, \mathbf{Li}_{k_m} \circ V_m)$ with P being a polynomial and $V_1, \ldots, V_m$ being rational functions; and $k_1, \ldots, k_m$ are integers satisfying $1 \leq k_i < n$ for every $i \in \underline{m}$.

Of relevance for web geometry are the webs defined by the functions U_i appearing in (6.19). The presence of a non-vanishing right-hand side $\mathbf{Elem}_n$ is an apparent obstruction to interpret (6.19) as an abelian relation of the web defined by the functions U_i. This difficulty can be bypassed because the classical polylogarithms have univalued 'cousins', denoted by $\mathcal{L}_n$, globally defined on $\mathbb{P}^1$ which satisfy, globally on $\mathbb{P}^2$, homogeneous analogues of every equation of the form (6.19) locally satisfied by $\mathbf{Li}_n$.

For $n \geq 2$, the **n-th modified polylogarithm** is the function

$$\mathcal{L}_n(z) = \Re_m\left(\sum_{k=0}^{n-1} \frac{2^k B_k}{k!} \log|z|^k\,\mathbf{Li}_{n-k}(z)\right),$$

defined for $z \in \mathbb{C} \setminus \{0, 1\}$.[6] It can be shown that these functions are well-defined real analytic functions on $\mathbb{C} \setminus \{0, 1\}$. They can be continuously extended to the whole projective line $\mathbb{P}^1$ by setting $\mathcal{L}_n(0) = 0$, $\mathcal{L}_n(\infty) = 0$, and $\mathcal{L}_n(1) = \zeta(n)$ for n odd, $\mathcal{L}_n(1) = 0$ for n even.

The existence of univalued versions of polylogarithms has been established by several authors. The ones introduced here have the peculiarity of satisfying *clean* versions of the functional equations for the classical polylogarithms presented above.

Theorem 6.5.1. *Let $n \geq 2$. The following assertions are equivalent:*

(a) *there exists a simply connected open subset of $\mathbb{P}^2$ where the functional equation $\sum_{i=1}^{N} c_i \, \mathbf{Li}_n(U_i) = \mathbf{Elem}_n$ holds true;*

(b) *the expression $\sum_{i=1}^{N} c_i \log |U_i|^{n-k} \mathcal{L}_k(U_i)$ is constant on $\mathbb{P}^2$ for $k = 2, \ldots, n$.*

For a proof of this result the reader is redirected to [98] and [37, pp. 45–46]. It implies that the web $\mathcal{W}_{\mathscr{E}_n}$ associated with a polylogarithmic relation $\mathscr{E}_n$ as (6.19) admits *polylogarithmic abelian relations* and hence is susceptible of having high rank.

For instance, according to Theorem 6.5.1, the function $\mathcal{L}_3$ verifies on $\mathbb{P}^2$ the homogeneous version of Spence–Kummer equation. For every $x, y \in \mathbb{R}$, the following identity holds true

$$
2\mathcal{L}_3(x) + 2\mathcal{L}_3(y) - \mathcal{L}_3\left(\frac{x}{y}\right) + 2\mathcal{L}_3\left(\frac{1-y}{1-x}\right)
$$
$$
+ 2\mathcal{L}_3\left(\frac{x(1-y)}{y(1-x)}\right) - \mathcal{L}_3(xy) + 2\mathcal{L}_3\left(-\frac{x(1-y)}{1-x}\right)
$$
$$
+ 2\mathcal{L}_3\left(-\frac{1-y}{y(1-x)}\right) - \mathcal{L}_3\left(\frac{x(1-y)^2}{y(1-x)^2}\right) = \frac{\zeta(3)}{2} .
$$

This equation can be complexified and the result, after differentiation, gives rise to an abelian relation for the **Spence–Kummer web** $\mathcal{W}_{SK}$ defined as

$$
\mathcal{W}_{SK} = \mathcal{W}\left(x, y, xy, \frac{x}{y}, \frac{1-x}{1-y}, \frac{x(1-y)}{y(1-x)}, \frac{x(1-y)}{(1-x)}, \frac{(1-y)}{y(1-x)}, \frac{x(1-y)^2}{y(1-x)^2}\right).
$$

This web seems to be to Spence–Kummer equation for the trilogarithm, what Bol's web is to Abel's equation for the dilogarithm. It was Hénaut in [75] who recognized it as a good candidate for being an exceptional 9-web. This was later settled independently by the second author [109] and Robert [119]. It has to be emphasized that back then in 2001, $\mathcal{W}_{SK}$ was the first example of planar exceptional

---

[6]In this definition, $\mathfrak{R}_m$ stands for the real part if n is odd and for the imaginary part otherwise; B_k is the k-th Bernoulli number: $B_0 = 1$, $B_1 = -1/2$, $B_2 = 1/6$, etc.

web to come to light after Bol's exceptional 5-web. Between the appearance of the two examples a hiatus of more or less 70 years took place.

One might think that all the webs naturally associated with the equations of the form (6.19) satisfied by the polylogarithms are exceptional (see [82, pp. 196–197]). Although these webs are certainly of high rank, they are not necessarily of the highest rank. For example, using Mihăileanu criterion, one can show by brute force computation that the webs associated with Kummer equations $\mathcal{K}(4)$ and $\mathcal{K}(5)$ are not of maximal rank (for details, see [107, Chap. VII]).

Nevertheless, it seems to exist a (large?) class of global exceptional webs with abelian relations expressed in terms of a natural generalization of the classical polylogarithms: the iterated integrals of logarithmic 1-forms on $\mathbb{P}^1$. For instance, all the elements of the family of 10-webs with parameters $a, b \in \mathbb{C} \setminus \{0, 1\}$

$$
\mathcal{W}_{a,b} = \mathcal{W}\left(x, y, \frac{x}{y}, \frac{1-y}{1-x}, \frac{b-y}{a-x}, \frac{x(1-y)}{y(1-x)}, \frac{x(b-y)}{y(a-x)}, \right.
$$
$$
\left. \frac{(1-x)(b-y)}{(1-y)(a-x)}, \frac{(bx-ay)(1-y)}{(y-x)(b-y)}, \frac{(bx-ay)(1-x)}{(y-x)(a-x)} \right)
$$

are exceptional webs. Through a method proposed by Robert in [119], it is possible to determine $\mathcal{A}(\mathcal{W}_{a,b})$ for no matter which a and b (see [112]) and deduce the maximality of the rank.

6.5.2 Series of Exceptional Webs

It is hard to imagine examples of webs simpler than the ones presented below.

$$
\mathcal{W}_1 = \mathcal{W}\left(x, y, x+y, x-y, xy\right)
$$
$$
\mathcal{W}_2 = \mathcal{W}\left(x, y, x+y, x-y, xy, x/y\right)
$$
$$
\mathcal{W}_3 = \mathcal{W}\left(x, y, x+y, x-y, x/y, x^2+y^2\right)
$$
$$
\mathcal{W}_4 = \mathcal{W}\left(x, y, x+y, x-y, xy, x^2+y^2\right)
$$
$$
\mathcal{W}_5 = \mathcal{W}\left(x, y, x+y, x-y, xy, x^2-y^2\right)
$$
$$
\mathcal{W}_6 = \mathcal{W}\left(x, y, x+y, x-y, xy, x/y, x^2-y^2\right)
$$
$$
\mathcal{W}_7 = \mathcal{W}\left(x, y, x+y, x-y, xy, x/y, x^2+y^2\right)
$$
$$
\mathcal{W}_8 = \mathcal{W}\left(x, y, x+y, x-y, xy, x^2-y^2, x^2+y^2\right)
$$
$$
\mathcal{W}_9 = \mathcal{W}\left(x, y, x+y, x-y, xy, x/y, x^2-y^2, x^2+y^2\right)
$$

It turns out that they all are exceptional as proved in [107, Appendice]. Notice that the webs $\mathcal{W}_1$ and $\mathcal{W}_2$ mentioned above are nothing more than the webs $\mathcal{A}_{III}^2$ and $\mathcal{A}_{IV}^2$ from Sect. 6.4.2. Moreover, $\mathcal{W}_3$ is equivalent to $\mathcal{A}_{II}^4$ under a linear change of coordinates. In the graph below the inclusions between them are schematically represented (by arrows).

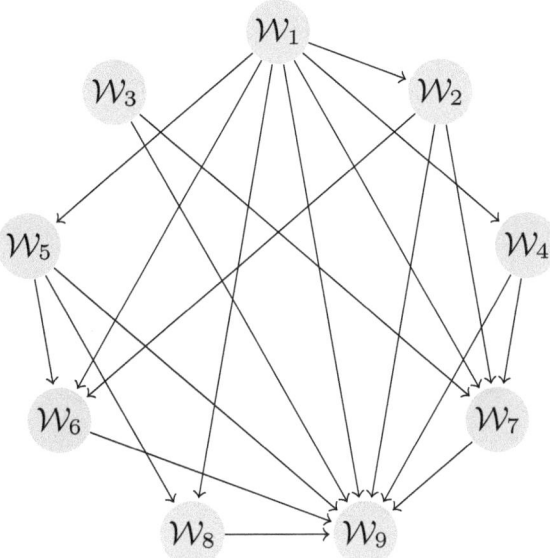

6.5.3 An Exceptional 11-Web

Let $\mathcal{F}_2$ be the degree two foliation on $\check{\mathbb{P}}^2$ induced by the rational 1-form $\check{y}(\check{y}-1)d\check{x} - \check{x}(\check{x}-1)d\check{y}$. It is nothing else than the pencil of conics $\frac{\check{x}(\check{y}-1)}{\check{y}(\check{x}-1)} = $ cte. Let C be the completely decomposable curve of degree 9 in $\check{\mathbb{P}}^2$ cut out by the homogeneous polynomial equation

$$\check{x}\,\check{y}\,\check{z}\,(\check{x}-\check{z})(\check{y}-\check{z})(\check{x}-\check{y})(\check{x}+\check{y})(\check{x}-\check{y}-\check{z})(\check{x}-\check{y}+\check{z}) = 0.$$

As can be seen above, C is the reunion of six lines invariant by $\mathcal{F}_2$ with three extra lines synthetically described as the lines joining the three singular points of the fibers of the pencil: these latter are cut out by $\check{x}+\check{y}$, $\check{x}-\check{y}-\check{z}$ and $\check{x}-\check{y}+\check{z}$.

The algebraic web $\mathcal{W}_C$ is formed by nine pencil of lines. If $\mathcal{W}_{\mathcal{F}_2}$ is the dual web of $\mathcal{F}_2$, in the sense of Sect. 1.4.3 of Chap. 1, then $\mathcal{W}_{\mathcal{F}_2} \boxtimes \mathcal{W}_C$ is an exceptional 11-web on $\mathbb{P}^2$. After a two-fold ramified covering it can be written as the completely decomposable web $\mathcal{W} = \mathcal{W}(F_1, \ldots, F_{11})$ where $F_1, \ldots, F_{11}$ are the rational functions below:

$$F_1 = \frac{(x-1)y}{(y-1)x}$$

$$F_2 = \left(\frac{y-x-1}{y-x+1}\right)^2 F_1$$

$$F_3 = \frac{(y-1)y}{(x-1)x}$$

$$F_4 = \frac{(y-x)y}{x-1}$$

$$F_5 = \frac{(x-1)y}{(y-1)x}$$

$$F_6 = \frac{(x-y+1)y}{x}$$

$$F_7 = \frac{x+y-1}{xy}$$

$$F_8 = \frac{(y-x-1)x}{y}$$

$$F_9 = \left(\frac{x-y+1}{y-x+1}\right) F_1$$

$$F_{10} = \frac{y(x-1)(x-y+1)}{x(y^2-xy-x+1)}$$

$$F_{11} = \frac{x(y-1)(y-x+1)}{y(x^2-xy-y+1)}.$$

Using Abel's method, the abelian relations of $\mathcal{W}$ can be explicitly determined. As a by-product, it follows that not only $\mathcal{W}$ is exceptional, but a certain number of its subwebs are exceptional as well. A partial list is provided by the following

Proposition 6.5.2. *The following ascending chain of subwebs of* $\mathcal{W}$

$$\mathcal{W}(F_1,\ldots,F_5) \subset \mathcal{W}(F_1,\ldots,F_6) \subset \ldots \subset \mathcal{W}(F_1,\ldots,F_{11}) = \mathcal{W}$$

is formed by exceptional webs.

It was David Marín together with the first author who guessed that this 11-web was interesting in what concerns its rank. The second author confirmed this intuition, proving the proposition above by using Abel's method.

6.5.4   Terracini's and Buzano's Webs

As explained in Sect. 4.3.4 of Chap. 4, there is a germ of smooth surface $S_{\mathcal{W}} \subset \mathbb{P}^5$ invariantly attached to every exceptional 5-web $\mathcal{W}$: the image of its Poincaré's map. Moreover, the geometry of $S_{\mathcal{W}}$ has rather special geometrical features as recalled below.

1. At a generic point, the second osculating space of $S_{\mathcal{W}}$ coincides with the whole $\mathbb{P}^5$;
2. The union of the tangent planes of $S_{\mathcal{W}}$ along one of the leaves of its Segre's web is included in a hyperplane;
3. The image of $\mathcal{W}$ by Poincaré's map of $\mathcal{W}$ is Segre's web of $S_{\mathcal{W}}$.

If $S \subset \mathbb{P}^5$ is a germ of surface in $\mathbb{P}^5$ satisfying the two conditions (1) and (2) above, it is natural to ask if its Segre's web $\mathcal{W}_S$, as defined in Sect. 1.4.4

of Chap. 1, is of maximal rank or not. A positive answer would establish an equivalence between the classification problem for exceptional 5-webs with a problem of projective differential geometry: the classification of surfaces subject to the constraints enumerated above. It is this problem which motivated Terracini and subsequently Buzano toward the results recalled below.

A surface $S \subset \mathbb{P}^5$ will be called an **exceptional surface** if

- it is not included in a Veronese surface;
- conditions (1) and (2) above are satisfied;
- its Segre's 5-web $\mathcal{W}_S$ is generically smooth.

Under this assumption, one proves the existence of five germs of curve $C_{S,i} \subset \check{\mathbb{P}}^5$ called the **Poincaré–Blaschke's curves** of S, satisfying

$$S = \bigcap_{i=1}^{5} (C_{S,i})^*$$

where $C^* \subset \mathbb{P}^5$ stands for the dual variety of a germ of curve $C \subset \check{\mathbb{P}}^5$. In other words, C^* is the subset of $\mathbb{P}^5$ corresponding to the hyperplanes $H \in \check{\mathbb{P}}^5$ tangent to C.

In [128], Terracini obtained a characterization of exceptional surfaces as solutions of a certain non-linear differential system. Under additional simplifying hypotheses, he succeeded to integrate explicitly the resulting system, and in this way proved the following result.

Theorem 6.5.3. *Up to projective automorphism, there are exactly four exceptional surfaces $S \subset \mathbb{P}^5$ for which three of their Poincaré–Blaschke curves—say $C_{S,i}$ for $i = 1, 2, 3$—are planar and such that the three associated planes $\langle C_{S,i} \rangle \subset \check{\mathbb{P}}^5$ have one point in common. One of these surface is the image of Poincaré's map of Bol's web, and the other three are the image of Poincaré's map of the following webs:*

$$\mathrm{Terr}(b) = \mathcal{W}\left(x, y, x+y, x-y, x^2-y^2\right)$$

$$\mathrm{Terr}(c) = \mathcal{W}\left(x, y, \frac{(x+y)^2}{1+y^2}, \frac{y(x^2y-2x-y)}{1+y^2}, \frac{x^2y-2x-y}{x^2+2xy-1}\right)$$

$$\mathrm{Terr}(d) = \mathcal{W}\left(x, y, x+y, \frac{x}{y}, \frac{x}{y}(x+y)\right). \tag{6.20}$$

Using Terracini's approach, Buzano [23] proved the following result.

Theorem 6.5.4. *Up to projective automorphism, there are exactly two exceptional surfaces in $\mathbb{P}^5$ for which three of their Poincaré–Blaschke curves—say $C_{S,i}$ for $i = 1, 2, 3$—are planar and satisfy*

(a) *for every distinct $i, j \in \underline{3}$, $\langle C_{S,i}, C_{S,j} \rangle \subset \check{\mathbb{P}}^5$ is a hyperplane;*
(b) *the intersection $\langle C_{S,1} \rangle \cap \langle C_{S,2} \rangle \cap \langle C_{S,3} \rangle$ is empty.*

They are the image of Poincaré's map of the following webs:

$$\text{Buz}(a) = \mathcal{W}\big(x, y, x + y, x - y, \tanh(x)\tanh(y)\big) \tag{6.21}$$

$$\text{Buz}(b) = \mathcal{W}\big(x, y, x + y, x - y, e^x + e^y\big).$$

It turns out that the webs (6.20) and (6.21) are all exceptional. Curiously, this was not proved by Terracini nor by Buzano. They focused on the differential-geometric problem. The exceptionality of these webs has been established just recently in [108, 113] (see also [107]), by using Abel's method.

Certain exceptional surfaces are transcendent, as, for example, the image of Poincaré's map of Bol's 5-web, while other are algebraic and even rational as the one associated with Terracini's web $\text{Terr}(b)$. Indeed, it can be verified that the image of Poincaré's map of $\text{Terr}(b)$ is the image of the following polynomial map:

$$(x, y) \longmapsto \left[1 : x^3 + y^3 : x^3 - y^3 : x^2 + y^2 : x^2 - y^2 : \big(x^2 - y^2\big)^2 \right].$$

A toy problem that might shed some light on the subject consists in determining the linear systems $\mathscr{L} \subset |\mathcal{O}_{\mathbb{P}^2}(q)|$, for small q, of dimension 5 for which the Zariski closure of the image of the associated rational map $\mathbb{P}^2 \dashrightarrow \mathbb{P}^5$ are exceptional surfaces. Notice that for $q = 4$, $\text{Terr}(b)$ is an example, and that 4 is the minimal q which can happen. Indeed for $q = 2$ one obtains a Veronese surface and for $q = 3$, the hyperplane containing the tangent spaces of leaves of Segre's web would pull-back to a cubic containing an irreducible component with multiplicity two. This implies that the pull-back of Segre's web to $\mathbb{P}^2$ is a linear, and consequently algebraic, web.

Appendix
On the History of Web Geometry

It seems impossible to grasp the ins and outs of a mathematical field without setting it back in its historical context. An attempt for web geometry, certainly incomplete and biased, is made in the next few pages.

Origins

If the *birth* of web geometry can be ascribed to the middle of the 1930s in Hamburg (see below), some precedents can be found as early as the middle of the nineteenth century. The concepts and problems of web geometry spring from two different fields of the nineteenth century mathematics: projective differential geometry and nomography.

Web geometry comes mainly from the first. At that time, projective differential geometry mainly consisted in the study of projective properties of curves and surfaces in the ordinary space $\mathbb{R}^3$, that is of their differential properties that are invariant up to homographies.

Gaussian geometry, which had appeared before, studied the properties of curves and surfaces in ordinary euclidean space that are invariant up to isometric transformations. Gauss and other mathematicians pointed out how useful the first and second fundamental forms are for the study of surfaces. They also brought to light the relevance of derived concepts, such as the principal, asymptotic, and conjugated directions. When considering the integral curves of these tangent direction fields, the mathematicians of the time were considering what they called 2-nets of '*lines*' on surfaces, that is the data of two families of curves, or in more modern terms, 2-webs. When they endeavored to generalize these constructions to projective differential geometry, some *3-nets* projectively attached to surfaces in $\mathbb{R}^3$ quite naturally made their appearance (for instance, Darboux introduced a 3-web called after him in [41]; see also Sect. 1.4.4 in this book).

© Springer International Publishing Switzerland 2015
J.V. Pereira, L. Pirio, *An Invitation to Web Geometry*, IMPA Monographs 2,
DOI 10.1007/978-3-319-14562-4

These webs were useful back then because they encoded properties of the surfaces under study. Thomsen's paper [129] is a good illustration of this fact. In this article, Thomsen shows that a surface in $\mathbb{R}^3$ is isothermally asymptotic[1] if and only if its Darboux 3-web is hexagonal.[2] At that time, the study of 3-webs on surfaces from the point of view of projective differential geometry was on the agenda.

A particular feature of Thomsen's result is his characterization of the geometric-differential property of being isothermally asymptotic by a closedness property of more topological nature that is (or not) verified by a configuration traced on the surface itself. It is this feature which struck some mathematicians and led to the study of webs at the beginning of the 1930s.

§

The second source of web geometry is nomography. This discipline, nowadays practically extinct, belonged to the field of applied mathematics in the 1900s. It was established as an autonomous mathematical discipline by M. d'Ocagne. It consisted in a method of 'graphical calculus' which allowed engineers to calculate rather fast. To explain its principle (which to-day appears rather naïve), let

$$L(V_1, V_2, V_3) = 0$$

be a mathematical law linking three physical variables V_1, V_2, and V_3. Is there a quick and accurate way to determine one variable, say V_i, from the other two: V_j and V_k? To solve this problem, people used nomograms. A nomogram is a graphic which represents curves according to values of the variables V_1, V_2, and V_3. For instance, to find the value of v_1 in function of values v_2 and v_3 of the variables V_2 and V_3 (respectively), one has to find the intersection point of the curves $V_2 = v_2$ and $V_3 = v_3$. Through (or near) this point goes a curve $V_1 = v_1$, and v_1 is the sought value (Fig. A.1).

What nowadays seems to be far from actual mathematics was once an important part of the mathematical culture. It was probably after considering some results of nomography that Hilbert formulated the 13th of the famous 23 problems that he stated at the International Congress of Mathematics of 1900.

The main disadvantage of nomography was the problem of its readability. Of course, the nomograms where the curves coincided with (pieces of) lines were easier to use. Hence the problem to know whether it is possible to linearize the curves of a given nomogram. Or equivalently, whether it is possible to linearize a 3-web of curves on the plane. For more precisions on the links between nomography and web geometry, the interested reader can consult [2].

---

[1]Geometers of the nineteenth century had established a very rich "bestiary" of surfaces in $\mathbb{R}^3$. The *isothermally asymptotic surfaces* (or "*F-surfaces*") formed one of the classes in their classification (see [51] for a modern definition.)

[2]Thomsen's result applies to real surfaces in $\mathbb{R}^3$, thus his statement is different depending on one takes place at a neighborhood of an elliptic point or a hyperbolic point of the considered surface.

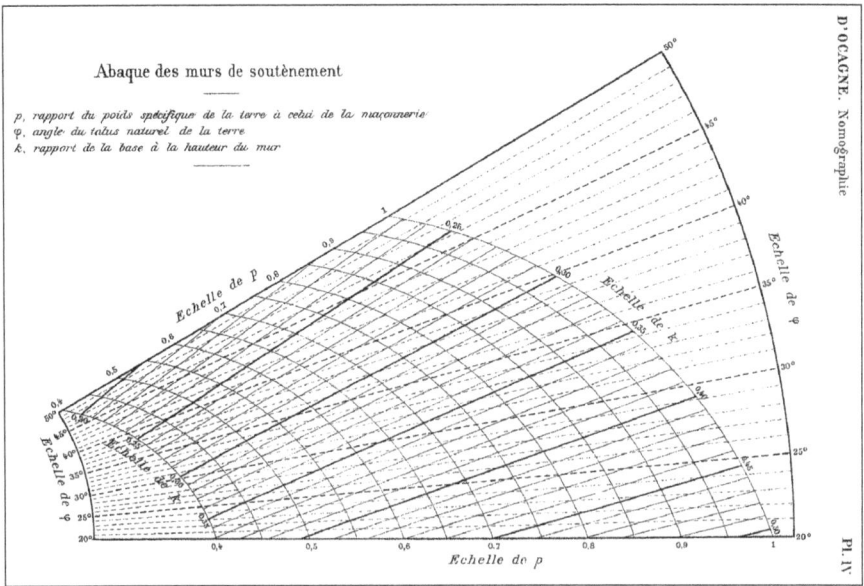

Fig. A.1 A nomogram from the book [43] by D'Ocagne

Birth of Web Geometry: Spring of 1927 in Naples

Thomsen's paper [129] is considered as marking the birth of web geometry. According to Blaschke (see the beginning of the foreword in [17]) this paper is the result of his Spring walks with Thomsen on Posillipo hill, in the vicinity of Naples, in 1927. Even if it concerns the study of some surfaces in $\mathbb{R}^3$, it clearly shows that a plane configuration made of three families of curves (i.e., a 3-web) admits local analytic invariants. It seems that the equivalence between the vanishing of the curvature of a 3-web (which is a condition of analytic nature) and the hexagonality condition (which is a property of topological nature)[3] struck these two mathematicians and led them (with others) to study the matter (Fig. A.2).

Early Developments: Hamburg School (1927–1938)

A short time after Thomsen's paper [129] was published, a group led by Blaschke was set up in Hamburg to do research on webs. Blaschke and his coworkers[4] found many results which established web geometry as a discipline. It is a remarkable fact

---

[3] See Theorem 1.2.4 in this book.

[4] Bol, Chern, Mayrhoffer, Podehl, Walberer were active members of this group. Kähler, Zariski, Reidemester, and others also worked on this subject but in an occasional way.

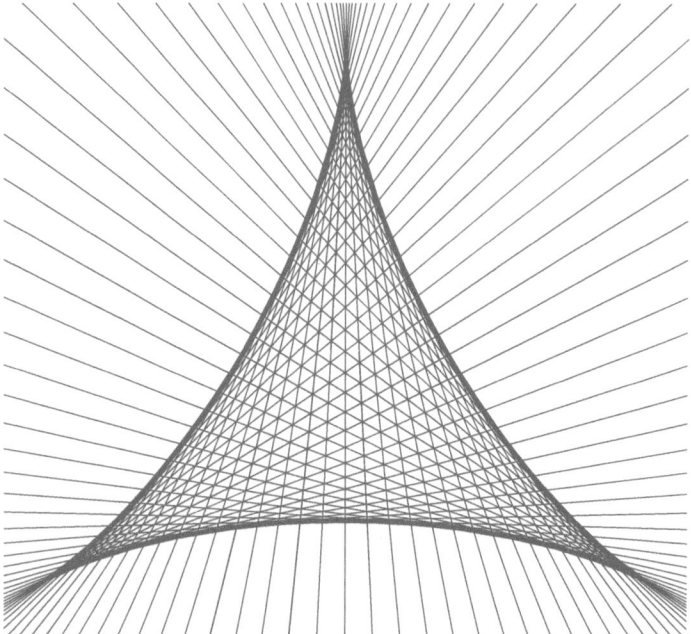

Fig. A.2 A 3-web with vanishing curvature formed by the tangents of a hypocycloid with 3 cusps

that a series of more than 60 papers were published in a variety of journals between 1927 and 1938, mainly by members of the Hamburg school of web geometry, under the common label of "Topologie Fragen der Differentialgeometrie."[5]

Their work developed along three main directions:

- The study of webs from the differential geometry viewpoint, through the analytical invariants which can be associated with them;
- The study of the relations between webs and abstract geometric configurations linked to the algebraic theory of (quasi-)groups;
- The interpretation of web geometry as a relative of projective algebraic geometry, notably via the notion of abelian relation.

Due to a lack of competence, we will not expand on the two former issues but we will focus on the latter direction which is also the main subject of this book. It deals with the links between webs and algebraic geometry, which have their origins in results obtained by Blaschke, Bol, and Howe.

At the beginning, these results were mainly about planar webs. Firstly, Blaschke came up with an interpretation of a theorem by Graf and Sauer [61] in the framework of web geometry. This theorem says that a linear 3-web carrying an abelian relation

---

[5]In English, "Topological questions of differential geometry".

is constituted by the tangents to a plane algebraic curve of class 3.[6] As soon as 1932, Blaschke and Howe [18] generalized this theorem to the case of linear k-webs carrying at least one complete abelian relation, thus bringing to light the usefulness of the notion of abelian relation. Bol's result giving the explicit bound $\frac{1}{2}(k-1)(k-2)$ on the dimension of the space of abelian relations of a planar k-web appeared shortly afterwards in [19] and allowed to define the rank of a web. Using this formalism, Howe noticed that Lie's result about the double-translation surfaces can be understood in the framework of web geometry as the striking fact that a planar 4-web of rank 3 is algebraizable. The relationship between the planar webs of maximal rank and Abel's Theorem was reported the following year by Blaschke in [15], which brought up the final definition of the notion of algebraic web. In the same paper, Blaschke expounded the generalization of Lie's Theorem to the 5-webs of maximal rank, a result which was later proved to be incorrect by Bol. Surprisingly, Blaschke also exhibited Bol's 5-web $\mathcal{B}_5$ as an example of non algebraizable 5-web of rank 5, while it is of maximal rank 6 (see below).

In 1933, Blaschke set about studying webs in dimension three. In [14], he established a bound $\pi(3,k)$ on the rank of a k-web of hypersurfaces in $\mathbb{C}^3$. One year later, Bol gave in [20] one of the most important results obtained at the time: *for $k \geq 6$, a k-web of hypersurfaces on $\mathbb{C}^3$ of maximal rank $\pi(3,k)$ is algebraizable.* This success certainly played a role in Blaschke's attempt in [15, 16] to obtain algebraization results for planar webs of maximal rank. Only in 1936 it was made clear that the result he was looking for was unattainable. In [21], Bol realized that $\mathcal{B}_5$ carries one more abelian relation, related to Abel's five terms equation for the dilogarithm; hence, it is an instance of a 5-web of rank 6 which is not algebraizable.

In this same year 1936, Chern defended his PhD dissertation on webs, written under Blaschke's direction. He then published two papers. The 60th issue of the "Topologische Fragen der Differentialgeometrie" series [30] is of special interest here. Generalizing Blaschke's result, he obtains a bound on the rank of a web of codimension one in arbitrary dimension[7] which now bears his name.

Thus, in 1936, most of the notions studied in this book had been brought to light. A general survey of the state the art at the time can be found in the third part of the book [17], to which the reader is asked to refer.

Finally, in this very year Blaschke shifted his interest from web geometry to integral geometry. Few members of the Hamburg school worked again on webs, with the notable exceptions of Blaschke and Chern, but this time in a different way (see below).

---

[6] By definition, the **class** of a reduced plane algebraic curve C is the number of tangent lines to C passing through a generic point of $\mathbb{P}^2$. Via projective duality, it corresponds to the degree of the dual curve $\check{C} \subset \check{\mathbb{P}}^2$.

[7] See Theorem 2.3.8 in this book.

Web Geometry in Mid Twentieth Century (1938–1960)

Blaschke strongly supported exchanges between mathematicians. From 1927 to 1960 he travelled a lot and had the opportunity to give lectures about web geometry in numerous countries (for instance, in Romania, Greece, Spain, Italy, the USA, India, Japan), thus inspiring people of various nationalities and backgrounds to do research on web geometry.[8]

It seems that Blaschke went to Italy many times during this period. As a by-product, an Italian school of web geometry developed at that time. Bompiani, Terracini, and Buzano were its most prominent contributors. Their work was chiefly about the links between the geometry of planar webs and the projective differential geometry of surfaces. In the 1950s and 1960s, a second Italian school of web geometry appeared, probably thanks to Bompiani's influence. He, Vaona, and Villa (among others) published papers on the projective deformation of planar 3-webs, but with no major outcome.

The work of the Romanian mathematicians Pantazi and Mihăileanu must be mentioned as well. During the 1930s and 1940s, they obtained interesting results on how to determine the rank of planar webs. These results were published as short notes in Romanian journals (see [87, 101]) and were then forgotten.

The war and later Blaschke's political stance during the war (see [123, p. 423]) put an end to his influence for some time. When things came back to normal, he gave lectures again, on webs among other things. Although he didn't obtain new results, these lectures induced new researches once more, for instance by Dou in Barcelona [44–46] and by Ozkän in Turkey [100].

Russian School (From 1965 Onwards)

More than at the Hamburg school, it is at the Moscow school of differential geometry that the Russian school of web geometry, led first by Akivis, and then by Akivis and Goldberg, seems to have its origin. Under the influence of the work of Élie Cartan, a Russian school of differential geometry developed in USSR at the instigation of Finikov from the 1940s onwards. Projective differential geometry was studied in full generality and involved the study of some nets (which could be called webs but only in a weak sense) projectively attached to (analytic) projective subvarieties. It is probably this fact which led to the study of webs for their own sake in arbitrary dimension and/or codimension, from the 1960s onwards. Akivis was joined by Goldberg quite early. They explored several directions in web geometry, published many papers and had many students.

---

[8]For instance, it is in attending some conferences given by Blaschke at Pekin in 1933 that Chern became interested in web geometry and decided to go and study at Hamburg.

The work of this school chiefly dealt with the differential geometry of webs and with the interactions between webs and the theory of quasigroups. The links with algebraic geometry were not their major concern. Their results had little influence in the West for two main reasons: (1) their papers were in Russian, hence they were not distributed in the West; (2) the method they used was the *Cartan–Laptev method*,[9] which non-specialists do not understand easily.

The reader who wishes to get an outline of the methods and results of this school may consult the books [3] and [59].

Chern's and Griffiths' Work (1977–1980)

Throughout his professional life Chern kept being interested in webs, particularly in the notion of web of maximal rank, as shown in [31, 32] and [33]. This point can be illustrated by quoting the last lines of [33]:

> *Due to my background I like algebraic manipulation, as Griffiths once observed. Local differential geometry calls for such works. But good local theorems are difficult to come by. The problem on maximal rank webs discussed above[10] is clearly an important problem, and will receive my attention.*
> *My mathematical education goes on.*

In 1978, he resumed working on webs of maximal rank jointly with Griffiths. In the long paper [34], they set about demonstrating that a k-web of codimension one and of maximal rank $\pi(n, k)$ is algebraizable when $n > 2$ and $k \geq 2n$. Their proof is not complete (cf. [36]) and it is necessary to make an extra non-natural assumption to ensure the validity of the result. They also got a sharp bound for the rank of webs of codimension two in [35].

Griffiths' interest in the subject probably came from the links between web geometry and algebraization results like the converse of Abel's Theorem discussed in Chap. 4. He discussed web geometry at the opening talk he gave at Abel's bicentennial conference held at Oslo in 2002, see [64].

Although it contains a non-trivial mistake, the paper [34] has been quite influential in web geometry. It has popularized the subject and led the Russian school to pay attention to the notions of abelian relations and rank. It is probably from [34] that Trépreau has taken up Bol's method to obtain a proof of the result originally aimed at by Chern and Griffiths. The present book would not exist if [34] had not been written. The reader should read it, as it contains a masterfully written introductory part putting things in perspective and offers different proofs of many of the results included here.

---

[9]Cartan–Laptev method is a reinterpretation/generalization of the methods of the mobile frame and of equivalence of É. Cartan by the Russian geometer G. Laptev.

[10]Chern is referring to the classification of webs of maximal rank.

Modern Developments (1980–2000)

A number of new interesting results in web geometry have been obtained in the last 20 years of the twentieth century. The only results mentioned here are those relating to rank, abelian relations and webs of maximal rank.

The abelian relations of Bol's web all come (after analytic prolongation) from its dilogarithmic abelian relation, which thus appears more fundamental than the other relations. In 1982, in [54], Gelfand and MacPherson found a beautiful geometric interpretation of this relation. In it Bol's web appears defined on the space of projective configurations of five points of $\mathbb{RP}^2$. In [40], Damiano considers, for $n \geq 2$ a curvilinear $(n + 3)$-web D_n naturally defined on the space of projective configurations of $n+3$ points in $\mathbb{RP}^n$. He shows that this web is of maximal rank and gives a geometric interpretation of the "main abelian relation" of D_n, thus obtaining a family of exceptional webs which generalizes Bol's web.

During the 1980s, Little studied some algebro-geometrical consequences of the converse to Abel's Theorem. He worked in particular on double-translation hypersurfaces (see for instance [85]) and published one paper on webs [86]. In it, he used some results of Mumford and Roitman on the space of 0-cycles on a projective surface of positive genus to construct examples of non-algebraizable two-codimensional webs of maximal rank from K3 surfaces embedded in projective space.

From 1980 to 2000, Goldberg studied the webs of codimension strictly bigger than one from the point of view of their rank. He obtained many results, most of which are expounded in [59]. Among his results, one finds the construction of several non-algebraizable webs of maximal rank.

Motivated by some questions in the geometric study of differential equations, Nakai began to study webs starting from the end of the 1980s using methods coming from differential topology. He worked mainly on webs in a real setting in relation with PDEs but he also produced papers on complex and real analytic webs such as [92, 93]. He also produced several expository papers [94–96] that certainly helped to make the subject more popular.

At the beginning of the 1990s, Hénaut started studying webs in the complex analytical realm. He published about 15 papers on the matter. His research is mainly about rank and abelian relations, and is concerned with webs of arbitrary codimension as well as planar webs (see [75] for an outline of the results he obtained before 2000). The papers [72, 73] have to be mentioned, as related with the topic of this book. At the time when he started working, the field attracted little attention. Without any doubt his tenacity played a major role to popularize web geometry in France, and in other countries as well.

At the occasion of the visit of Nakai, some *Journées sur les tissus* were organized at Toulouse in 1996. These 'journées' gathered researchers from different backgrounds, as testified by the content of the book [66].

To finish this very quick overview of the works on webs during the 1980s and the 1990s, let's mention some papers by researchers working on foliations and dynamical systems such as [28, 56].

Recent Developments (Since 2000)

The activity on webs during the two decades before 2000 announced an even greater vitality in this area for the years to follow. In particular, major results concerning the problem of algebraizing webs of maximal rank have been obtained since 2001. We mention below those that we consider the most important.

In 2001, the second author [107, 109] and Robert [119] independently showed that the Spence-Kummer 9-web associated with the trilogarithm is an example of exceptional planar web. This was the second example of such a web, the other one having being discovered by Bol more than 60 years before.

A notable event for the field was the conference *Géométrie des tissus et équations différentielles* co-organized by Hénaut and Nakai. This conference, held at the CIRM in 2003, was attended by researchers from all over the world. For some mathematicians it was their first opportunity to meet web geometry. In the 10 years following this conference, a myriad of new examples of planar exceptional webs were brought to light [91, 106, 108, 113] showing that the classification problem for non-algebraic planar webs of maximal rank is more difficult than suggested by Chern and Griffiths, but also more interesting due to the variety and distinct natures of known webs of this kind.

In 2005, Trépreau provided a proof of the result which Chern and Griffiths aimed at in [34], i.e. the algebraization of one-codimensional k-webs on $(\mathbb{C}^n, 0)$ of maximal rank, when $k \geq 2n$ and $n > 2$. This result being one of the main subjects of this book, we will not comment it any longer here.

The results hitherto mentioned were the sign that the old problematic of the algebraization of webs, that goes back to the Hamburg school of web geometry, was undergoing a revival, as is testified by the Bourbaki seminar [105] devoted to this topic.

From 2005, Trépreau and the second author of this book started working on the algebraization of maximal rank webs of codimension strictly greater than 1. If Poincaré's approach generalizes and is still fruitful to study these webs, it leads to a geometrical problem that does not really hold for webs of codimension 1. Indeed, using constructions similar to the ones presented in this book, it is possible to reduce the question of the algebraization of maximal rank webs of codimension bigger than 1 to the classification of some extremal projective varieties carrying a 'big' family of rational normal curves.

This question has been completely solved in [114] in all but one particular case, for which some unexpected exceptions arise. Using the classifications obtained in [114], the authors were able to obtain a general algebraization result for maximal rank webs of codimension greater than 2 in [115]. The result is the expected one (namely '*maximal rank webs are algebraizable*') except for one unexpected case corresponding to the geometric exceptions mentioned just above.

Finally, this particular case has been investigated in [111] where the author constructed '*algebraic exceptional webs*', that is webs of maximal rank that are algebraic in a generalized sense, but not in the classical one.

§

So, as for maximal rank webs of codimension 1, for which exceptional webs exist in dimension 2, there are algebraic exceptional webs of codimension bigger than 2. All these exceptional webs, planar or not, are far from being well understood. Their number and their still mysterious geometry make them exciting mathematical objects worth to be studied further. We hope that they will.

To conclude, it is worth mentioning that although major results about the algebraization of webs have been obtained in the last decade, the field must not be reduced to the study of exceptional webs. Many other natural questions remain widely open. Among them, let us mention the one about the algebraization of webs of codimension $c \geq 2$ with maximal q-rank when $q < c$ (see [76] for details). Another one is linked to the nice but seemingly forgotten result by Blaschke and Walberer [17, §37] about the algebraization of maximal rank webs whose codimension does not divide the dimension of the ambient space.

There is still a lot of space for future research in web geometry.

Bibliography

1. Abel, N.H.: Méthode générale pour trouver des functions d'une seule quantité variable lorsqu'une propriété de ces fonctions est exprimée par une équation entre deux variables. In: Œuvres complètes de N.H. Abel, Tome 1, pp. 1–10. Grondhal Son, Rhode Island (1981)
2. Aczél, J.: Quasigroups, nets, and nomograms. Adv. Math. **1**, 383–450 (1965). Doi:10.1016/0001-8708(65)90042-3
3. Akivis, A., Shelekhov, A.: Geometry and algebra of multidimensional three-webs. In: Mathematics and Its Applications, vol. 82. Kluwer, Dordrecht (1992)
4. Akivis, M., Goldberg, V.V., Lychagin, V.: Linearizability of d-webs, $d \geq 4$, on two-dimensional manifolds. Sel. Math. **10**, 431–451 (2004). Doi:10.1007/s00029-004-0362-x
5. Andreotti, A.: Théorèmes de dépendance algébrique sur les espaces complexes pseudo-concaves. Bull. Soc. Math. France **91**, 1–38 (1963). http://www.numdam.org/item?id= BSMF_1963__91__1_0
6. Aluffi, P., Faber, C.: Plane curves with small linear orbits II. Int. J. Math. **11**, 591–608 (2000). Doi:10.1142/S0129167X00000301
7. Arbarello, E., Cornalba, M., Griffiths, P.A., Harris, J.: Geometry of Algebraic Curves, vol. I. Grundlehren der Mathematischen Wissenschaften, vol. 267. Springer, New York (1985)
8. Arnol'd, V.I.: Geometrical methods in the theory of ordinary differential equations. In: Grundlehren der Mathematischen Wissenschaften, vol. 250. Springer, New York (1988)
9. Ballico, E.: The bound of the genus for reducible curves. Rend. Mat. Appl. **7**, 177–179 (1987)
10. Barlet, D.: Le faisceau $\omega_X^\bullet$ sur un espace analytique X de dimension pure. In: Norguet, F. (ed.) Fonctions de Plusieurs Variables Complexes III. Lecture Notes in Mathematics, vol. 670, pp. 187–204. Springer, Berlin (1978). Doi:10.1007/BFb0064400
11. Barth, W., Hulek, C., Peters, C., Van de Ven, A.: Compact Complex Surfaces. Springer, New York (2004)
12. Beauville, A.: Le problème de Schottky et la conjecture de Novikov. Séminaire Bourbaki, Vol. 1986/87. Astérisque No. 152–153, 101–112 (1987). http://eudml.org/doc/110074
13. Beltrami, E.: Resoluzione del problema: riportari i punti di una superficie sopra un piano in modo che le linee geodetische vengano rappresentata da linee rette. Ann. Math. **1**, 185–204 (1865). Doi:10.1007/BF03198517
14. Blaschke, W.: Abzählungen für Kurvengewebe und Flächengewebe. Abh. Math. Hamburg Univ. **9**, 299–312 (1933). Doi:10.1007/BF02940656
15. Blaschke, W.: Textilegeometrie und abelsche integrale. Jber. Deutsch. Math.-Ver. **43**, 87–97 (1933)
16. Blaschke, W.: Über die Tangenten einer ebenen Kurve fünfter Klasse. Abh. Math. Hamburg Univ. **9**, 313–317 (1933). Doi:10.1007/BF02940657

© Springer International Publishing Switzerland 2015
J.V. Pereira, L. Pirio, *An Invitation to Web Geometry*, IMPA Monographs 2,
DOI 10.1007/978-3-319-14562-4

17. Blaschke, W., Bol, G.: Geometrie der Gewebe. Die Grundlehren der Math, vol. 49. Springer, Berlin (1938)
18. Blaschke, W., Howe, G.: Über die Tangenten einer ebenen algebraischen Kurve. Abh. Math. Hamburg Univ. **9**, 166–172 (1932). Doi:10.1007/BF02940641
19. Bol, G.: On n-webs of curves in a plane. Bull. Am. Math. Soc. **38**, 855–857 (1932). http://projecteuclid.org/euclid.bams/1183496400
20. Bol, G.: Flächengewebe im dreidimensionalen Raum. Abh. Math. Hamburg Univ. **10**, 119–134 (1934). Doi:10.1007/BF02940669
21. Bol, G.: Über ein bemerkenswertes Fünfgewebe in der Ebene. Abh. Math. Hamburg Univ. **11**, 387–393 (1936). Doi:10.1007/bf02940735
22. Bryant, R., Manno, G., Matveev, V.: A solution of a problem of Sophus Lie: normal forms of two-dimensional metrics admitting two projective vector fields. Math. Ann. **340**, 437–463 (2008). Doi:10.1007/s00208-007-0158-3
23. Buzano, P.: Determinazione e studio di superficie di S_5 le cui linee principali presentano una notevole particolarità. Ann. Math. Pura Appl. **18**, 51–76 (1939). Doi:10.1007/BF02413766
24. Buzano, P.: Tipi notevoli di 5-tessuti di curves piane. Boll. Unione Mat. Ital. **1**, 7–11 (1939)
25. Casale, G.: Feuilletages singuliers de codimension un, Groupoïde de Galois et intégrales premières. Ann. Inst. Fourier **56**, 735–779 (2006). Doi:10.5802/aif.2198
26. Cavalier, V., Lehmann, D.: Global stucture of webs in codimension one. Preprint arXiv:0803.2434v1 (2008)
27. Cavalier, V., Lehmann, D.: Ordinary webs of codimension one. Ann. Sci. Norm. Super. Pisa **11**, 197–214 (2012). Doi:10.2422/2036-2145.201003_007
28. Cerveau, D.: Théorèmes de type Fuchs pour les tissus feuilletés Astérisque No. **222**, 49–92 (1994)
29. Cerveau, D., Mattei, J.-F.: Formes intégrables holomorphes singulières. Astérisque, vol. 97. Société Mathematique de France, Paris (1982)
30. Chern, S.-S.: Abzählungen für Gewebe. Abh. Math. Hamburg Univ. **11**, 163–170 (1935). Doi:10.1007/BF02940720
31. Chern, S.-S.: Web geometry. Bull. Am. Math. Soc. **6**, 1–8 (1982). http://projecteuclid.org/euclid.bams/1183548587
32. Chern, S.-S.: The mathematical works of Wilhelm Blaschke—an update. In: Burau, W., Chern, S.-S. et al. (eds.) Wilhelm Blaschke Gesammelte Werke Band, vol. 5, pp. 21–23. Thales-Verlag, Essen (1985)
33. Chern, S.-S.: My mathematical education. In: Yau, S.-T. (ed.) Chern—A Great Geometer of the Twentieth Century, pp. 1–17. International Press, Hong Kong (1992)
34. Chern, S.-S., Griffiths, P.A.: Abel's theorem and webs. Jahresberichte der Deutsch. Math.-Ver. **80**, 13–110 (1978). http://eudml.org/doc/146681
35. Chern, S.-S., Griffiths, P.A.: An inequality for the rank of a web and webs of maximum rank. Ann. Sc. Norm. Super. Pisa **5**, 539–557 (1978). http://www.numdam.org/item?id=ASNSP_1978_4_5_3_539_0
36. Chern, S.-S., Griffiths, P.A.: Corrections and addenda to our paper:"Abel's theorem and webs". Jahresberichte der Deutsch. Math.-Ver. **83**, 78–83 (1981)
37. Colmez, P.: Arithmétique de la fonction zêta. In: Berline, N., Sabbah, C. (eds.) La fonction zêta, pp. 37–164. Éd. École Polytech, Palaiseau (2003)
38. Coxeter, H.: Introduction to Geometry. Reprint of the 1969 edition. Wiley Classics Library. Wiley, New York (1989)
39. Dalbec, J.: Multisymmetric functions. Beiträge Algebra Geom. **40**, 27–51 (1999). http://www.emis.de/journals/BAG/vol.40/no.1/3.html
40. Damiano, D.: Webs and characteristic forms of Grassmann manifolds. Am. J. Math. **105**, 1325–1345 (1983). Doi:10.2307/2374443
41. Darboux, G.: Sur le contact des courbes et des surfaces. Bull. Soc. Math. France **4**, 348–384 (1880). http://www.numdam.org/item?id=BSMA_1880_2_4_1_348_1
42. De Medeiros, A.: Singular foliations and differential p-forms. Ann. Fac. Sci. Toulouse Math. **9**, 451–466 (2000). http://eudml.org/doc/73521

43. D'Ocagne, M.: Nomographie: Les Calculs Usuels Effectués au Moyen des Abaques. Gauthier-Villars, Paris (1891)
44. Dou, A.: Plane four-webs. Mem. Real Acad. Ci. Art. Barcelona **31**, 133–218 (1953)
45. Dou, A.: Rang der ebenen 4-Gewebe. Abh. Math. Sem. Univ. Hamburg **19**, 149–157 (1955). Doi:10.1007/BF02988869
46. Dou, A.: The symmetric representation of hexagonal four-webs. Collect. Math. **9**, 41–58 (1957)
47. Eisenbud, D., Harris, J.: On varieties of minimal degree (a centennial account). Proc. Symp. Pure Math. **46**, 3–13 (1987). Doi:10.1090/pspum/046.1
48. Eisenbud, D., Green, M., Harris, J.: Cayley-Bacharach theorems and conjectures. Bull. Am. Math. Soc. **33**, 295–324 (1996). Doi:10.1090/S0273-0979-96-00666-0
49. El Haouzi, A.: Sur la torsion des courants $\bar{\partial}$-fermés sur un espace analytique complexe. C. R. Acad. Sci. Paris Sér. I Math. **332**, 205–208 (2001). Doi:10.1016/S0764-4442(00)01804-8
50. Fabre, B.: Nouvelles variations sur les théorèmes d'Abel et de Lie. Thèse de Doctorat de L'Université Paris VI, 2000. Available at http://tel.archives-ouvertes.fr/.
51. Ferapontov, E.: Integrable systems in projective differential geometry. Kyushu J. Math. **54**, 183–215 (2000). Doi:10.2206/kyushujm.54.183
52. Fubini, G., Čech, E.: Introduction à la Géométrie Projective Différentielle des Surfaces. Gauthier-Villars, Paris (1931).
53. Fuchs, D., Tabachnikov, S.: Mathematical Omnibus. Thirty Lectures on Classic Mathematics. American Mathematical Society, Providence (2007)
54. Gelfand, I., MacPherson, R.: Geometry in Grassmannians and a generalization of the dilogarithm. Adv. Math. **44**, 279–312 (1982). Doi:10.1016/0001-8708(82)90040-8
55. Gelfand, I., Kapranov, M., Zelevinsky, A.: Discriminants, resultants, and multidimensional determinants. Mathematics: Theory & Applications. Birkhäuser, Boston (1994)
56. Ghys, E.: Flots transversalement affines et tissus feuilletés. Mém. Soc. Math. France **46**, 123–150 (1991). http://www.numdam.org/item?id=MSMF_1991_2_46__123_0
57. Ghys, E.: Osculating curves. Slides of a talk (2007). http://www.umpa.ens-lyon.fr/~ghys/Publis.html
58. Godbillon, G.: Géométrie Différentielle et Mécanique Analytique. Hermann, Paris (1969)
59. Goldberg, V.V.: Theory of multicodimensional $(n + 1)$-webs. Mathematics and Its Applications, vol. 44. Kluwer, Dordrecht (1988)
60. Goldberg, V.V., Lychagin, V.: On the Blaschke conjecture for 3-webs. J. Geom. Anal. **16**, 69–115 (2006). Doi: 10.1007/BF02930988
61. Graf, H., Sauer, R.: Über dreifache Geradensysteme in der Ebene, welche Dreiecksnetze bilden. Akad. Math.-Naturwiss. Abt. 119–156 (1924)
62. Griffiths, P.A.: Variations on a theorem of Abel. Invent. Math. **35**, 321–390 (1976). Doi:10.1007/BF01390145
63. Griffiths, P.A., Harris, J.: Principles of algebraic geometry. Pure and Applied Mathematics. Wiley-Interscience, New York (1978)
64. Griffiths, P.A.: The legacy of Abel in algebraic geometry. In: Laudal, O., Piene, R. (eds.) The Legacy of Niels Henrik Abel, pp. 179–205. Springer, New York (2004)
65. Grifone, J., Muzsnay, Z., Saab, J.: On the linearizability of 3-webs. Proceedings of the Third World Congress of Nonlinear Analysts, Part 4 (Catania, 2000). Nonlinear Anal. **47**, 2643–2654 (2001). Doi:10.1016/S0362-546X(01)00385-6
66. Grifone, J., Salem, E. (eds.): Web Theory and Related Topics. World scientific, Singapore (2001)
67. Harris, J.: A bound on the geometric genus of projective varieties. Ann. Sc. Norm. Super. **8**, 35–68 (1981). http://www.numdam.org/item?id=ASNSP_1981_4_8_1_35_0
68. Harris, J.: Curves in projective space. With the collaboration of David Eisenbud. Séminaire de Mathématiques Supérieures, **85**, Presses de l'Université de Montréal, Montréal (1982)
69. Hartshorne, R.: Cohomological dimension of algebraic varieties. Ann. Math. **88**, 403–450 (1968). Doi:10.2307/1970720

70. Hartshorne, R.: Algebraic Geometry. Graduate Texts in Mathematics, vol. 52. Springer, New York (1977)

71. Hartshorne, R.: The genus of space curves. Ann. Univ. Ferrara Sez. **40**, 207–223 (1994). Doi:10.1007/BF02834521

72. Hénaut, A.: Sur la linéarisation des tissus de $\mathbb{C}^2$. Topology **32**, 531–542 (1993). Doi:10.1016/0040-9383(93)90004-F

73. Hénaut, A.: Caractérisation des tissus de $\mathbb{C}^2$ dont le rang est maximal et qui sont linéarisables. Compos. Math. **94**, 247–268 (1994). http://www.numdam.org/item?id=CM_1994__94_3_247_0

74. Hénaut, A.: Tissus linéaires et théorèmes d'algébrisation de type Abel-inverse et Reiss-inverse. Geom. Dedicata **65**, 89–101 (1997). Doi:10.1023/A:1004916502107

75. Hénaut, A.: Analytic web geometry. In: Grifone, J., Salem, E. (eds.) Web Theory and Related Topics, pp. 150–204. World Scientific, Singapore (2001)

76. Hénaut, A.: Formes différentielles abéliennes, bornes de Castelnuovo et géométrie des tissus. Comment. Math. Helv. **79**,25–57 (2004). Doi:10.1007/s00014-003-0787-4

77. Hénaut, A.: On planar web geometry through abelian relations and connections. Ann. Math. **159**, 425–445 (2004). Doi:10.4007/annals.2004.159.425

78. Hénaut, A.: Planar web geometry through abelian relations and singularities. In: Griffiths, P.A. (ed.) Inspired by Chern, Nankai Tracts in Mathematics, vol. 11, pp. 269–295. World Scientific, Singapore (2006)

79. Henkin, G., Passare, M.: Abelian differentials on singular varieties and variations on a theorem of Lie-Griffiths. Invent. Math. **135**, 297–328 (1999). Doi:10.1007/s002220050287

80. Kleiman, S.: What is Abel's theorem anyway? In: Laudal, O., Piene, R. (eds.) The Legacy of Niels Henrik Abel, pp. 395–440. Springer, New York (2004)

81. Lane, E.: A Treatise On Projective Differential Geometry. University of Chicago Press, Chicago (1942)

82. Laudal, O., Piene, R.: The Legacy of Niels Henrik Abel–the Abel Bicentennial, Oslo, 2002. Springer, New York (2004)

83. Lewin, L.: Polylogarithms and Associated Functions. North-Holland, New York-Amsterdam (1981)

84. Liouville, R.: Mémoire sur les invariants de certaines équations différentielles et sur leurs applications. J. de l'Éc. Polyt. Cah. LIX. 7–76 (1889)

85. Little, J.B.: Translation manifolds and the converse of Abel's theorem. Compos. Math. **49**, 147–171 (1983). http://www.numdam.org/item?id=CM_1983__49_2_147_0

86. Little, J.B.: On webs of maximum rank. Geom. Dedicata **31**, 19–35 (1989). Doi:10.1007/BF00184156

87. Mihăileanu, N.: Sur les tissus plans de première espèce. Bull. Math. Soc. Roum. Sci. **43**, 23–26 (1941)

88. Mumford, D.: The Red Book Of Varieties And Schemes. Lecture Notes in Mathematics, vol. 1358, Springer, New York (1999)

89. Muzsnay, Z.: On the problem of linearizability of a 3-web. Nonlinear Anal. **68**, 1595–1602 (2008). Doi:10.1016/j.na.2006.12.033

90. Marín, D., Pereira, J.V.: Rigid flat webs on the projective plane. Asian J. Math. **17**, 163–191 (2013). http://projecteuclid.org/euclid.ajm/1383923439

91. Marín, D., Pereira, J. V., Pirio, L.: On planar webs with infinitesimal automorphisms. In: Griffiths, P.A. (ed.) Inspired by Chern, Nankai Tracts in Mathematics vol. 11, pp. 351–364, World Scientific, Singapore (2006)

92. Nakai, I.: Topology of complex webs of codimension one and geometry of projective space curves. Topology **26**, 475–504 (1987). Doi:10.1016/0040-9383(87)90043-7

93. Nakai, I.: Curvature of curvilinear 4-webs and pencils of one forms: variation on a theorem of Poincaré, Mayrhofer and Reidemeister. Comment. Math. Helv. **73**, 177–205 (1998). Doi:10.1007/s000140050051

94. Nakai, I.: Web geometry: why does a Riemann surface come from a (double) translation surface? Sūrikaisekikenkyūsho Kōkyūroku **1065**, 163–177 (1998)

95. Nakai, I.: Web geometry and the equivalence problem of the first order partial differential equations. In: Grifone, J., Salem E. (eds.) Web Theory And Related Topics, pp. 150–204. World Scientific, Singapore (2001)

96. Nakai, I.: Web geometry of solutions of holonomic first order PDEs. Nat. Sci. Rep. Ochanomizu Univ. **53**, 107–110 (2002)

97. Nickalls, R., Dye, R.: The geometry of the discriminant of a polynomial. Math. Gaz. **80**, 279–285 (1996). Doi:10.1.1.190.9465

98. Œsterlé, J.: Polylogarithmes. Séminaire Bourbaki, Exp. No. 762, Astérisque No. **216**, 49–67 (1993)

99. Olver, P.: Equivalence, Invariants, and Symmetry. Cambridge University Press, Cambridge (1995)

100. Özkan, A.: Über die Sechseckbedingungen bei einer n-Kurvenwabe in der Ebene. Abh. Math. Hamburg Univ. **21**, 95–98 (1957). Doi:10.1007/BF02941929

101. Pantazi, A.: Sur la détermination du rang d'un tissu plan. C. R. Acad. Sci. Roum. **2**, 108–111 (1938)

102. Pareschi, G., Popa, M.: Castelnuovo theory and the geometric Schottky problem. J. Reine Angew. Math. **615**, 25–44 (2008). Doi:10.1515/CRELLE.2008.008

103. Pereira, J.V.: Vector fields, invariant varieties and linear systems. Ann. Inst. Fourier **51**, 1385–1405 (2001). Doi:10.5802/aif.1858

104. Pereira, J.V., Sanchez, P.F.: Transformation groups of holomorphic foliations. Comm. Anal. Geom. **10**, 1115–1123 (2002)

105. Pereira, J.V.: Algebraization of Codimension one Webs. Séminaire Bourbaki: Volume 2006/2007. Astérisque No. **317**, 243–268 (2008)

106. Pereira, J.V., Pirio, L.: The classification of exceptional CDQL webs on compact complex surfaces. Int. Math. Res. Not. **12**, 2169–2282 (2010). Doi:10.1093/imrn/rnp208

107. Pirio, L: Équations fonctionnelles abéliennes et géométrie des tissus. Thèse de Doctorat de l'Université Paris VI (2004). Available electronically at http://tel.archives-ouvertes.fr.

108. Pirio, L.: Sur les tissus plans de rang maximal et le problème de Chern. C. R. Math. Acad. Sci. **339** 131–136 (2004). Doi:10.1016/j.crma.2004.04.022

109. Pirio, L.: Abelian functional equations, planar web geometry and polylogarithms. Selecta Math. **11**, 453–489 (2005). Doi:10.1007/s00029-005-0012-y

110. Pirio, L.: Sur la linéarisation des tissus. L'Enseignement Mathématique **55**, 285–328 (2009). Doi:10.4171/LEM/55-3-5

111. Pirio L.: Tissus algébriques exceptionnels. Preprint arXiv:1305.6493 (2013)

112. Pirio, L., Robert, G.: Unpublished manuscript (2005)

113. Pirio, L., Trépreau, J.-M.: Tissus plans exceptionnels et fonctions Thêta. Ann. Inst. Fourier **55**, 2209–2237 (2005). Doi:10.5802/aif.2159

114. Pirio L., Trépreau J.-M.: Sur les variétés X dans $\mathbb{P}^N$ telles que par n points passe une courbe de X de degré donné. Bull. Soc. Math. France **141**, 131–196 (2013)

115. Pirio L., Trépreau J.-M.: Sur l'algébrisation des tissus de rang maximal. Int. Math. Res. Not. (2014). Doi:10.1093/imrn/rnu066

116. Pirola, G.P., Schlesinger, E.: Monodromy of projective curves. J. Algebraic Geom. **14**, 623–642 (2005). Doi:10.1090/S1056-3911-05-00408-X

117. Poincaré, H.: Sur les surfaces de translation et les fonctions abéliennes. Bull. Soc. Math. France **29**, 61–86 (1901)

118. Ripoll, O.: Géométrie des tissus du plan et équations différentielles. Thèse de Doctorat de l'Université Bordeaux 1 (2005). Available electronically at http://tel.archives-ouvertes.fr.

119. Robert, G.: Relations Fonctionnelles Polylogarithmiques et Tissus Plans. Prépublication, vol. 146. Université Bordeaux 1, Bordeaux (2002)

120. Ripoll, O.: Properties of the connection associated with planar webs and applications. Preprint arXiv:math.DG/0702321 (2007).

121. Robert, G.: Poincaré maps and Bol's Theorem. Available electronically at http://kyokan.ms.u-tokyo.ac.jp/~topology/GHC/data/Robert.pdf (2005).

122. Rosenlicht, M.: Equivalence relations on algebraic curves. Ann. Math. **56**, 169–191 (1952). Doi:10.2307/1969773
123. Segal, S.: Mathematicians Under the Nazis. Princeton University Press, Princeton (2003)
124. Segre, C.: Le linee principali di una superficie di S5 e una proprietà caratteristica della superficie di Veronese. Atti R. Acc. Lincei XXX, 200–203/227–231 (1921)
125. Spencer, D.: Overdetermined systems of linear partial differential equations. Bull. Am. Math. Soc. **75**, 179–239 (1969). Doi:10.1090/S0002-9904-1969-12129-4
126. Tabachnikov, S., Timorin, V.: Variations on the Tait-Kneser theorem. Preprint arXiv:math/0602317 (2006)
127. Tedeschi, G.: The genus of reduced space curves. Rend. Sem. Mat. Univ. Politec. Torino **56** 81–88 (1998)
128. Terracini, A.: Su una possibile particolarità delle linee principali di una superficie. I i II. Atti Accad. Naz. Lincei **26**, 84–91/153–158 (1937)
129. Thomsen, G.: Un teorema topologico sulle schiere di curve e una caratterizzazione geometrica sulle superficie isotermo-asintotiche. Boll. Un. Mat. Ital. Bologna **6**, 80–85 (1927)
130. Trépreau, J.-M.: Algébrisation des Tissus de Codimension 1 – La généralisation d'un Théorème de Bol. In: Griffiths, P.A. (ed.) Inspired by Chern, Nankai Tracts in Mathematics, vol. 11, pp. 399–433. World Scientific, Singapore (2006)
131. Tresse, A.: Détermination des invariants ponctuels de l'équation différentielle ordinaire du second ordre $y'' = \omega(x, y, y')$. Leipzig. 87 S. gr. 8°. (1896)
132. Wang, J. S.: On the Gronwall conjecture. J. Geom. Anal. **22**, 38–73 (2012). Doi:10.1007/s12220-010-9184-6
133. Weimann, M.: Trace et calcul résiduel: une nouvelle version du théorème d'Abel inverse, formes abéliennes. Ann. Fac. Sci. Toulouse Math. **16**, 397–424 (2007). Doi:10.5802/afst.1154
134. Wirtinger, W.: Lie's Translationmannigfaltigkeiten und das Abelsche Integrale. Monatsch. Math. Phys. **46**, 384–431 (1938). Doi:10.1007/BF01792693
135. Wood, J.: A simple criterion for local hypersurfaces to be algebraic. Duke Math. J. **51**, 235–237 (1984). Doi:10.1215/S0012-7094-84-05112-3

Index

© Springer International Publishing Switzerland 2015
J.V. Pereira, L. Pirio, *An Invitation to Web Geometry*, IMPA Monographs 2,
DOI 10.1007/978-3-319-14562-4